Our Way or the Highway

Our Way or the Highway

Inside the Minnehaha Free State

Mary Losure

University of Minnesota Press
Minneapolis • London

MINNESOTA

The University of Minnesota Press gratefully acknowledges the generous assistance
provided for the publication of this book by the Margaret W. Harmon Fund.

Maps in this book were created by Parrot Graphics.

Published by the University of Minnesota Press
111 Third Avenue South, Suite 290
Minneapolis, MN 55401-2520
http://www.upress.umn.edu

Printed in the United States of America on acid-free paper

Library of Congress Cataloging-in-Publication Data

Losure, Mary.
    Our way or the highway : inside the Minnehaha Free State / Mary Losure.
        p. cm.
      ISBN 0-8166-3905-1 (PB : alk. paper)
    1. Deep ecology—Minnesota. 2. Highways—Environmental aspects—
Minnesota.   I. Title.
    GE198.M5 L68 2002
    333.7'2'09776—dc21

For Don

Contents

Preface   ix

1. **The Camp**   1

2. **Operation Coldsnap**   25

3. **The Raid**   39

4. **Little Crow's Children**   67

5. **"I'll Do Anything"**   95

6. **Piestruck**   125

7. **Spring Comes to the Minnehaha Spiritual Encampment**   145

8. **Tree by Tree**   155

9. **Four Bur Oaks**   195

10. **Coldwater Nation**   209

Afterword   227

Acknowledgments   229

Note on Sources   231

Preface

At four o'clock in the morning, the squad cars streamed north. A river of headlights flowed for more than a mile along old Highway 55, a narrow, potholed stretch between the Twin Cities International Airport and downtown Minneapolis. City police, county sheriffs, and state troopers from across Minnesota had gathered in secret on the airport tarmac that night for the largest combined law enforcement operation in the state's history. It was December 20, 1998. Their destination was a squatters' camp in a working-class south Minneapolis neighborhood. The camp's inhabitants called it the Minnehaha Free State.

Born of an improbable coalition of scraggly young people, Native Americans, and a sixty-seven-year-old woman willing to try anything to stop a highway from taking her home, the protest camp had grown and flourished for more than six months. Its roots went back much further, to a neighborhood's decades-long fight against a freeway.

I covered the story of the protest camp as an environmental reporter for Minnesota Public Radio, beginning in the late summer of 1998. At first, the struggle against the widening and rerouting of Highway 55 seemed like a purely local story, less important than many other environmental issues I'd covered for Minnesota Public Radio and National Public Radio. But gradually, almost in spite of myself, I was drawn to it.

Its characters seemed to have stepped out of a novel. A grandmother befriends members of the radical environmental group Earth First! and appears in the national *Earth First! Journal* in full camouflage, fist raised. A burly, part-Dakota pipe fitter with a "Born Again Savage" bumper sticker on his van abandons his old life to defend what he believes are sacred sites of his Dakota ancestors. A dreamy, dreadlocked New Age seeker hitchhikes to Minneapolis and spends days communing with a doomed cottonwood tree in the path of the highway project. A Minneapolis police lieutenant with a Harvard education watches the trees fall in his neighborhood and ponders the nature of activism.

As the oddball, unpredictable tale of the Minnehaha Free State unfolded, it raised troubling questions about mainstream America. How do we make the decisions that shape the future of our cities? What binds us into communities? What, in the twenty-first century, do we hold sacred?

And after covering the protest against Highway 55 for more than a year, I realized that it was not a purely local story. It was one toenail of a much larger animal. We mainstream reporters find this animal hard to understand, much less describe. For lack of a better label, we might call it an "activist subculture" or an "underground movement," but that's not it, exactly. It's more like a shared spirit.

People at the Minnehaha Free State sometimes called it "the resistance," but even they found it hard to describe, and they didn't like labels anyway. It's not an organization or even a movement, they said. It has no official leaders or spokespersons, and no membership rolls. But its followers tend to share certain core beliefs. They oppose globalization and biotechnology. They support sustainable agriculture. They worry about the tide of cement engulfing the planet.

Who are they? Sometimes they are young people with nose rings and green hair, or self-described "old lefties" with middle-aged spread and Birkenstocks, but not always. Sometimes they drive late-model cars and live in the suburbs.

However difficult to define, this movement, this spirit of resistance, is out there. When the time and place are right, it surfaces. It showed itself in Seattle in November 1999, when fifty thousand people appeared as if from nowhere to protest the meeting of the World Trade Organization. It surfaced again in antiglobalization protests in Washington, D.C., and in Quebec City. Because I had seen the same spirit in action fifteen minutes from where I live, I didn't wonder where the multitudes in Seattle came from or who organized them. I knew a lot about them already. I had met their brothers and sisters at the Minnehaha Free State.

The Highway 55 "protesters," as we in the media called them, had shortcomings like everyone else. They could be strident and irrational. They could be naive and impractical. Some stretched the truth. Others seemed to do everything possible to alienate the ordinary citizens they hoped so fervently would join their cause. Often they seemed to lack any clear strategy. But the better I came to know them, the more respect I had for what they were trying to do. As the struggle in Minneapolis continued month after month, as people camped in the snow in the brutal cold of two Minnesota winters, I finally began to realize what they had known all along: it was about more than just a road.

CHAPTER 1

The Camp

The four oaks stood at the edge of a weedy field in south Minneapolis. It was a place the burgeoning city had somehow left behind, an area not much bigger than a block or two, where asphalt ended in tall grass, woods, and a scattering of wildflowers. For more than a hundred years, the oaks had sprouted softly lobed leaves, tiny as squirrels' ears, each spring. In the brief Minnesota summers, the leaves glowed deeper green in the sunlight; in autumn, they dried to a tough, leathery brown. As the long cold approached, the trees dropped acorns, each small cap fringed with the bristles that give bur oaks their name. In winter, the oaks showed their bare bones: large, heavy branches in a rounded crown, the thick, gray-brown bark gouged with darker streaks flowing like riverbeds down their trunks to deep taproots. Around them, the city surged outward, bigger and noisier each year, but the oaks grew silently, millimeter by millimeter, one ring a season.

Undisturbed, the oaks might have stood for another century. But in the summer of 1998, it seemed they had only a season or two left in their long lives. They stood in the path of a highway.

That August, a Minneapolis pipe fitter named Jim Anderson stepped from the sunlight into the oaks' shade and froze, listening.

Over the din of traffic and the roar of airplanes, he heard a steady chanting—voices rising and falling in a Dakota song.

"I went in between the south and the east tree and I heard *HEY, yay yay yay yay*," Jim remembered. "I said, 'Did anybody hear that?'" Jim's cousin and uncle stared at him. "It was something that never ever happened to me before in my life. I got goose bumps. I don't know if that was a religious experience or what. If that's what they call it, I guess I had one then."

Jim, his cousin Michael, and his uncle Bob had come to see the trees because Chris Leith, a respected Dakota elder, had told them the grove was a sacred place.

Leith was a spiritual adviser and healer for the Mdewakanton branch of the Dakota Nation. He lived near the Prairie Island reservation, about an hour's drive south along the Mississippi River from Minneapolis. For thirty years, Leith had been a Sundance chief, an honor given only to those who work with the spirits through dreams and visions. In recent years, he had been Jim's teacher. "I have a lot of respect for Chris Leith. When he told me that it was a sacred place of our people, and that we had to try to get them to not destroy it, I thought right then that we had to do that," Jim said.

The four oaks stood in a diamond with points north, south, east, and west. Leith told Jim they had been used as Dakota burial scaffolds. The diamond stood near the steep wooded gorge of the Mississippi, on bluffs where Dakota villages once clustered. A mile to the southeast, the Mississippi joined the Minnesota River. There soldiers had built Fort Snelling, where hundreds of Dakota men, women, and children had been held prisoners after the Dakota Uprising of 1862. No one knows what happened to the bodies of the Dakota who died there, but Leith thought some might be buried in the field near the four oaks. And just a few minutes' walk away, hidden behind a security fence on the abandoned campus of the Federal Bureau of Mines, a spring flowed all year round. The water was sacred, Leith said.

But if it was up to Jim Anderson to save a Dakota sacred place, he was at first glance an unlikely candidate for the job. For one thing, he didn't look particularly Indian. The son of a Swedish American father and a part-Dakota mother, he sometimes joked that he was "Swindian." With his prominent, European-looking nose and a plain Midwestern accent that held not a trace of Dakota cadence, he had lived most of his life as a white person.

But by the summer day when Jim stepped beneath the four oaks, he had been studying with Chris Leith for three years, learning the Dakota language. Jim attended sweat lodge and Sundance ceremonies often. He was "trying to find the way," as he put it. He had always worn his hair long; now he put on a silver eagle ring, a feather-shaped earring, and T-shirts with Native American images. He didn't care if people called him a wannabe. "Wannabe—who wouldn't want to be once you find out what Native spirituality actually is and how simple and basic and down to earth and common sense it is?" he asked. "How couldn't you want to be?"

Jim heard the voices again a few days later, when he was hiking with his cousin along the bluffs of the Mississippi River near the four trees. "I've never heard anything more than that," Jim said. "Other people are more attuned to spirituality than I am, I think, and they see spirits, or some people do. A lot of the elders do, and I haven't. But I felt that presence at that time, and I heard that chanting. I knew that there was something there."

Then one night, Jim checked in to his pipe fitter's job at the University of Minnesota. "I went to the tool crib to get some tools, and it's a big stainless steel countertop there. And the guy's behind the countertop. And I ordered tools, what I needed, and he went to get them. And I looked down and there's this little skull there," Jim said. "It was bleached just white and I'm looking at it, and he came back and he goes, 'Where'd you get that?' And I said it was laying right here on the counter. He goes, 'Oh no it wasn't, there was nothing on the counter.'

"Right away I just got this big rush," Jim said. He asked what the thing was, and the man said it was a squirrel's skull. "I asked Chris Leith about it. He said the Great Spirit or *Wakan Tonka* will give you some gifts sometimes, to make sure you don't think you're going crazy or something. So to me, that's what it was. Later that day after I had found out, I couldn't hardly even grab it, I was shaking so bad.

"A lot of strange things that people can't explain do happen in this world, and now I realize that," Jim said. Chris Leith told him it was a sign. "He said that's the squirrel nation or the tree nation that's asking for help."

Jim told his boss he needed time off. He was too agitated to come to work. A week went by, and he lost his job, but it didn't matter. He had more important things to do.

Jim and other highway opponents took their stand in a south Minneapolis neighborhood with streets named for the poet Longfellow's vanished, romantic Indians: Hiawatha, Minnehaha, Nokomis, Nawadaha. Highway 55, also called Hiawatha Avenue, ran through it, south from downtown Minneapolis to the airport. In those days, the highway's southern stretch was narrow and potholed. Riverview Road, where the protest began, was just off the old highway, directly in the path of the planned reroute. It was a quiet, leafy street only one block long, set on the western bluff of the Mississippi River.

Despite the noise of the city—the traffic on Highway 55, the bleeps of trucks backing up, the constant roar of jets from the nearby airport—Riverview Road seemed like a throwback to a quieter, greener time. Modest one- and two-story houses lined a street shaded by old oaks along the edge of a gorge that dropped steeply to the Mississippi. At night, raccoons and foxes came out of the woods and wandered through the yards. In the summertime, crickets chirped from the underbrush.

At its south end, the street's shade gave way to vacant lots and a parklike expanse of lawns dotted with spreading oaks. Beyond lay an open field: a patchwork of bare earth crisscrossed by tire treadmarks and bordered by unkempt grasses and cottonwood seedlings. On one side of the field, a path led through a patch of forest to the four bur oaks where Jim heard the voices.

Plans in the Minnesota Department of Transportation offices showed the Highway 55 reroute drawn dead center through the four oaks. Highway 55 would be widened and swung along the banks of the Mississippi River, through the field, the four oaks, and the houses on one side of Riverview Road.

Minneapolis pipe fitter Jim Anderson believed the Highway 55 reroute would destroy sites sacred to his Dakota ancestors. Photograph copyright Keri Pickett.

For years, MnDOT, as it was known, had been buying and razing homes on the side of Riverview Road closest to the Mississippi. (The houses on the other side of the street would be left intact when the road went through.) The foundations of houses that had already been bulldozed were still visible in the vacant lots at the end of the street. By the spring of 1998, seven houses still stood in the way of the highway. MnDOT had purchased six of them, and their owners had moved away. But the family in the seventh house refused to sell.

Al and Carol Kratz had lived on Riverview Road for forty-three years. They had raised their son and daughter there. Al was a retired printer. Carol had worked as a stenographer, but mostly she had stayed home to raise the children. She was sixty-seven years old, the grandmother of four. She refused to leave her one-and-a-half-story house, with its trellis of morning glories near the front door, its hedge of old-fashioned bridal wreath bushes, and its backyard shaded by the big oaks along the Mississippi.

A small, energetic woman with a short, practical haircut, Carol moved as easily as a teenager. She dressed in tennis shoes and sweatshirts, and seemed like someone you would see at a yard sale or a mall, buying Beanie Babies for her grandchildren. But she had been fighting the highway for a long time.

Carol and others in south Minneapolis had formed a group called the Park and River Alliance, which argued that rerouting Highway 55 along the river would destroy century-old oaks on the edge of one of Minnesota's oldest and most beloved parks, Minnehaha Park, which lay just north of Riverview Road. They collected seven thousand signatures opposing the reroute. They tried to oust the mayor and a city council candidate who declined to take a stand against the project. They hired a lawyer and filed an environmental lawsuit.

But nothing seemed to work. The mayor and city council

Carol Kratz, a sixty-seven-year-old grandmother who lived in the path of the reroute, enlisted the help of the radical environmental group Earth First! to fight the project. This photograph of Carol in front of her house appeared in the national *Earth First! Journal* under the headline "Calling All Squatters." Photograph courtesy of *Earth First! Journal*.

candidate won. A judge threw out the lawsuit, ruling that it had been filed too late. Then, at a rally in the spring of 1998, Carol met some activists from Earth First!, a radical environmental group.

The Earth First!ers had spent parts of the last two winters in an old canvas Girl Scout tent in the deep snow of northern Minnesota, trying to stop the logging of a stand of century-old pines. In the end, some of the trees had been saved and the rest were logged. Now the group was ready for a new campaign. Carol was thrilled. "I thought, 'Is this ever wonderful!'" she said later.

Soon half a dozen Earth First!ers set up their Girl Scout tent on Carol's front lawn, under two huge spruce trees. Carol ran an extension cord out from her house. They called it Camp Two Pines, after the Kratz family's joking name for the spot where her brother used to park his R.V. when he came to visit. The *Earth First! Journal*, the movement's national magazine, ran an article, headlined "Calling All Squatters," about the camp. A photo showed a small figure stationed in front of the tent, beneath a sign reading "Camp Two Pines." Carol wore dark sunglasses and full camouflage, and she raised her clenched fist high.

One of Camp Two Pines' first occupants was a young Earth First! activist who went by the name of Tumbleweed. He was a short, slight twenty-two-year-old with shoulder-length blond hair and a bushy beard that seemed too big for his thin face. He had a springy step and boundless optimism. He belonged to a Chicago subgroup of Earth First! and had learned about the proposed reroute of Highway 55 at Earth First!'s annual Midwest Spring Rendezvous. On the last day of the gathering, a rain-soaked camp-out in an isolated park in southern Minnesota, a Minneapolis activist named Susu Jeffrey had driven down from the Twin Cities to speak. She told several dozen soggy Earth First!ers about the neighborhood's long struggle against the Highway 55 reroute. "I remember hearing about this antiroad campaign and thinking, 'Wow! This is pretty intense,'" Tumbleweed said.

Earth First!'s militant and dramatic antiroad campaigns in Britain were legendary among its followers. Now, Tumbleweed thought, this would be the group's first major antiroad campaign in this country, right here in the Midwest. "I remember this feeling of, my God, we're going to create a monster," he said later. "It would be unlike any campaign the U.S has ever seen."

Tumbleweed had watched videos of Earth First! protests against a road known as M-11 in east London. "[The M-11 campaign] had this huge, huge outpouring of support. Anarchists, community members, Green Party, people like that," Tumbleweed said. "They occupied houses, and they locked [themselves] down. And it wasn't just Earth First! kind of people, it was regular, everyday, active citizens realizing they had to take nonviolent direct action." He hoped the same thing would happen in Minnesota.

All that spring, the Earth First!ers took shifts sleeping in the tent. They handed out anti–Highway 55 literature and waited. They knew the Minnesota Department of Transportation planned to raze the six empty houses on either side of Carol's that summer. Somebody called up the demolition company and asked when. The person on the phone said August 10.

That day, at four o'clock in the morning, three women barricaded themselves in a room in one of the vacant houses. Near another house, a man who called himself Joe Hill, after the early-twentieth-century songwriter and organizer for the Industrial Workers of the World, sat in a seat suspended twenty feet above the ground from the legs of a steel tripod. He had chained his neck to one of the legs, putting himself at risk of injury or death if the tripod was suddenly toppled. Other activists chained their wrists to cement pads, then they all waited for the police to come and saw them out.

"We figured it would be a symbolic action," Tumbleweed said later. "We'd get arrested after a day or two and that would be it." But to their surprise, the police let them stay, and the bulldozers never came. Instead, that same day Jim Anderson and his cousin Michael heard about the protest, showed up, and pitched their tipi near the Earth First! tent. "The synchronicity of it was amazing," Tumbleweed said. "It was meant to be."

August 10, 1998, marked the beginning of what the activists came to call the Minnehaha Free State, an encampment that grew with each passing week. "There was this feeling in the air of *we're going to win this*," Tumbleweed said. "For once we're going to win, and there's nothing they can do."

Meaghan, a nineteen-year-old canvasser for the Sierra Club's Minneapolis office, learned about the camp a few days later when somebody brought in a newspaper article. A photograph showed Joe Hill in his tripod, locked down by the neck. Meaghan had never heard of Earth First! although she had always considered herself an environmentalist. She and a friend decided to check it out.

Meaghan is thoughtful and articulate, a quick study who had always taken an interest in the world around her, but she would come to think of that first visit to the camp as the beginning of her political education. The camp was full of "crazy weirdos," she later said jokingly, but there were also older, experienced activists. "I started learning all these amazing things and talking with all these really knowledgeable people." She discovered alternative media sources she had never heard of while she was growing up in Minnetonka, an affluent Minneapolis suburb. "I started realizing how I'd been lied to all my life. A lightbulb went on in my head.

An activist sits suspended from the legs of a steel tripod in the path of the reroute. Photograph copyright 2001 *Star Tribune*/Minneapolis–St. Paul.

"On Monday, I didn't go back to work. I went to the Free State," Meaghan recalled. She realized that the Sierra Club was not for her; it was too mainstream, too full of people willing to compromise for political gain, too amenable to saving one piece of forest by sacrificing another. The Earth First! position was different: their motto was "Zero cut, no compromise."

"I couldn't really be anywhere else," she said. "I felt like it was too important for me to be there. The experience of being at the Free State—for me, it changed my whole life."

To picture Riverview Road after the camp began, imagine an average city block in an ordinary Midwestern town. On one side of the street, life goes on as usual; people mow their lawns and wash their cars and watch television. But on the other side, almost overnight, a circus has pitched its tents.

The longtime residents of Riverview Road opened their front doors each morning and looked across the street to a different world. The scent of wood smoke and burning sage drifted on the air. Dozens of fluorescent nylon dome tents and brightly painted tipis clustered under Carol's two big spruces and on the front lawns of the vacant houses next door. More dome tents sprouted like toadstools in the woods along the riverbank. In the crooks of the big oaks, Earth First!ers built wooden platforms that looked like children's tree houses. To get to Carol's front door, you had to weave through piles of firewood, card tables stacked with anti–Highway 55 literature, and a brick-lined fire pit dug in Carol's neatly mowed front lawn. Activists from the American Indian Movement joined the protest soon after it began, and they and Jim Anderson and his cousin Michael kept a ceremonial fire burning twenty-four hours a day. The AIM activists wore camouflage pants and red bandanna headbands. They threw cedar boughs, sweet grass, and tobacco on the fire.

Behind one of the vacant houses, they built a sweat lodge of bent willow limbs.

On sunny weekends, dreadlocked, nose-ringed young Earth First!ers in black T-shirts, cargo pants rolled up to the knee, and heavy hiking boots strolled on the lawns with gray-haired 1960s activists wearing dangly earrings, Birkenstocks, and long flowing skirts. The two generations chatted like a motley family gathered for a reunion. Dog walkers on their way to the river picked up literature at the card tables. Gawkers drove by in big American cars with their windows rolled up.

At night, songs and drumming carried across Riverview Road and disturbed the neighbors. Some complained to the city about the wood smoke and noise, and about people urinating in the bushes. One neighbor told Carol, "This is all your goddamned fault."

Retelling the story later, Carol asked innocently, "How could one person cause this?"

It certainly seems she did, her listener replied.

"I suppose," she admitted, laughing.

In Carol's house, a framed photograph taken that summer showed Carol and other Highway 55 activists standing in front of her house, smiling into the camera, arms entwined around each other's shoulders. "Tumbleweed," she said nostalgically, looking at his bearded face in the photo. "He's such a lovable kid. He's everywhere. Whatever's happening, he's there, hopping around like a little bunny rabbit. So sweet, so caring.

"They were wonderful neighbors, they were great," Carol said of the activists. "Every day, I'd get a hug. If I missed a day, I'd get two hugs."

Carol needed comforting. That summer, her husband, Al, was dying of lung disease and suffering the late stages of Alzheimer's disease. His face was sunken and gray. He couldn't be left alone for long. He wandered around the house at the end of his oxygen tube, muttering and groaning and whistling. He repeated snatches of conversation, changing words and improvising on them like a jazz musician. Sometimes he yelled obscenities.

Carol knew Al was terrified of being put in a nursing home. She wanted him to live out his life in the house they had shared for so long. After she put Al to bed in the evenings, she would step out her front door and sit with the young activists around the fire. It helped calm her nerves.

Most mornings, Meaghan woke up at four o'clock to go on security patrol. She walked the perimeter of the camp with a handheld radio, watching for police and strangers. Everyone took shifts.

Patrol cars cruised slowly by. Sometimes the officers got out and asked questions. Meaghan figured they were mostly just bored and glad to talk to anyone who was awake at that time of night, but sometimes it seemed to her they were truly curious about what the protesters were doing there and what they wanted.

At eight o'clock in the morning, when Meaghan's shift ended, a woman called Sunshine made breakfast at the Café Two Pines, a makeshift kitchen set up in the abandoned garage next to Carol's house. Meaghan took her turn washing dishes, or attended strategy meetings, or listened to music in the café, or gave camp tours to visitors.

Most weeks, an ancient, brightly painted former school bus belonging to Sister's Camelot, a free food distribution group, wheezed into camp bearing produce salvaged from organic wholesalers. As word of the camp spread, supporters brought food and clothing. Gradually, people moved into the six vacant houses on either side of Carol's. They gave each a name: North House, Kitchen House, ABC House, Tripod House, Pink House, and South House.

From his south Minneapolis apartment, an Earth First! organizer named Bob Greenberg faxed out a steady stream of press releases. "OCCUPATION: MINNEHAHA FREE STATE—DAY 7. EARTH FIRST! DIGS IN FOR THE LONG HAUL" he announced on August 16. "ACTIVISTS FROM ILLINOIS, WISCONSIN, AND IOWA JOIN IN THE RESISTANCE."

"MINNEHAHA LIBERATED AREAS HOLD OFF BULLDOZERS FOR FIRST MONTH!" said the September 9 fax. "MANY MORE ARE SURE TO FOLLOW!"

Reporters soon came to recognize Greenberg's style. A typical fax announced that Earth First! activists, Native Americans, and neighborhood groups had joined together to rally against "MnDOT's evil plans" and that "mass civil disobedience" was "only one of the aces that the protesters have up their sleeves."

"What planet are these people from?" one news editor asked after reading a particularly outlandish message.

The first few months of camp were an anxious, busy time for Jim Anderson. He bought a daily organizer, the first one he'd ever owned, and tried to keep track of the many meetings that now filled his days. The book's pages showed appointments with lawyers and journalists, and speaking engagements at schools, powwows, corn feeds, and ceremonies. He also kept a diary in a lined spiral notebook, the pages marked off from August 10. He drove his rattling van from one place to another without, it seemed, any clear plan of action. A sampling of the diary entries read: "Went home got back for meeting with Earth First! about getting along in camp; went to Chris Leith and had sweat; went to MnDOT gave papers to ask for new EIS [Environmental Impact Statement, an environmental study required for highways]; kicked out three people from bus for drinking but more for flopping; had water ceremony at sacred fire."

The hardest thing was being away from his four-year-old son, Joey, who lived with Jim's ex-wife but stayed with Jim's parents during the day. Before he had moved to camp, Jim worked nights, lived with his parents, and saw Joey every day. Now Jim saw his son only occasionally and noted it down each time: "Went home to Joey." But he was consumed by his mission.

That fall, Jim and a friend drove to Green Grass, a tiny settlement on the barren, rolling grasslands of the Cheyenne River Sioux Reservation in South Dakota. They wanted to see an elder, Arvol Looking Horse.

For generations, Arvol Looking Horse's family members had been spiritual leaders of the Lakota. The Lakota are one of three tribes, Lakota, Dakota, and Nakota, known collectively under the name Dakota, which means "allies." Arvol Looking Horse was the Keeper of the Sacred White Buffalo Calf Pipe, an honor recognized both on and off the reservation. When officials from the Museum of the South Dakota State Historical Society repatriated a medicine man's rattle acquired in 1906, they returned it to Arvol Looking Horse. The rattle had belonged to Elk Head, the ninth keeper of the sacred pipe. Museum officials noted in the *Federal Register*, "Mr. Arvol Looking Horse, keeper of the sacred pipe, traces his ancestry directly and without interruption to Elk Head, his great-great-great grandfather."

Arvol Looking Horse kept trailers and horses that he used for what he called Unity Rides—ceremonial rides between places sacred to the Dakota and other peoples. Jim hoped Looking Horse would agree to go on a ride to the four trees. "I brought my pipe to him and he accepted my pipe. And I told him what we were doing here, and he said it was really good," Jim said. "These people can look in your eyes and see who you are, and they know by how you talk where your heart is."

Arvol Looking Horse invited Jim to pray in the wooden shed where the Sacred White Buffalo Calf Pipe is kept. In November, Looking Horse brought eight horses to Birch Coulee, Minnesota, for a four-day ride to Minneapolis. They chose Birch Coulee because it is a sacred place to the Dakota, who fought there during the Dakota Uprising.

Looking Horse held ceremonies along the way. "It was very strong," Jim said. "I mean, when you get people like that, he's like the pope of the Dakota Nation, you could put it in that analogy." In one place, eagles flew over the ceremony. "They show us, those are our grandfathers, the spirits in the sky. To us, that's what happens when you're doing the right thing, you get the okay from the animal world too."

Looking Horse and six others rode their horses into Minneapolis across the freeway bridge that crosses the confluence of

Lakota spiritual leader Arvol Looking Horse (*front row, second from right*) leads a Unity Ride from Birch Coulee, Minnesota, to the protest camp. Staff photograph by *St. Paul Pioneer Press*; copyright *St. Paul Pioneer Press*.

the Mississippi and Minnesota Rivers at Mendota, about a mile southeast of Riverview Road. The riders wore full ceremonial garb—headdresses made of feathers or wolf skin, quill necklaces and breastplates—and held their ceremonial staffs high. The horses and riders clopped down the shoulder, protected from the traffic whizzing all around by a slow-moving caravan of cars and trucks with their lights on. At the end of the bridge, the procession moved placidly into the slow lane of the freeway. Surrounded by the caravan, the riders navigated the multilane interchange of Highway 62 and Highway 55 without incident, then followed Highway 55 toward the camp. No police accompanied them, since no one had requested a police escort or parade permit. Aside from a few motorists who honked and waved, few people in the city were aware of the event. At the camp, Arvol Looking Horse held a ceremony under the four trees.

At the end of Riverview Road, people from camp built a corral. "I can remember going down there and the moon was shining through the trees and the horses were making their little sounds. It was quite an experience to have horses in the city," Carol said. "We were violating another rule," she added, smiling at the memory.

The Earth First!ers in camp, meanwhile, had been keeping busy. At night, dressed in black and carrying shovels, they met behind the houses. They were building what are known as lockdown sites: places where activists can chain themselves down when the police come. The crews soon became specialists in a particular kind of lockdown known as a "dragon," which they had learned to make by reading how-to manuals.

"There are books out there," explained an Earth First!er named Bill Busse. Bill was a sturdy, bearlike man in his mid thirties, a longtime local activist who worked part-time jobs and devoted most of his time to working for causes he believed in. He lived "low on the capitalist ladder," as he put it, rather than "feeding the machine" by spending his life working for the system.

"We'd read about it in the *Earth First! Journal*, we'd read about it in a book called *Ecodefence* by [Earth First! cofounder] Dave Foreman, and then we'd read about it in a book called the *Earth First! Direct Action Manual*. We had a lot of time on our hands. Dragons were easy to make. It was easy to get the supplies. We got really good at it," Bill said candidly.

To build a dragon, you dig a hole about the size of a garbage can, then down the center you put a piece of black plastic pipe, just big enough for an arm to fit through. At the deep end of the pipe, you lay a piece of steel reinforcing rod. You fill the hole with cement, top it off with a steel fire door with an armhole cut in it, and the dragon is ready. If you plan to lock down to your dragon, you wear a chain around your wrist, bolted to a metal clip like the ones on dog leashes. When the time comes, you plunge your arm in up to the shoulder, clip your wrist chain to the steel rod deep inside the dragon, and wait. You can clip and unclip your wrist at will, but the only way anyone else can unhook you is to dig up the dragon.

The Earth First!ers built dragons not only in the yards but also in the basements of the vacant houses, cracking the cement floors with sledgehammers so they could dig the holes. They reinforced the room in the northernmost house—the one where the women had barricaded themselves on August 10—adding fifty-five-gallon drums filled with concrete. They even tried to dig a tunnel system under the houses. (Antiroad activists had dug such tunnels in Britain to thwart heavy equipment, which couldn't drive over them for fear of breaking through.) But on

Riverview Road the ground proved too sandy for tunneling. The attempt, Bill said, had been a "miserable failure."

In bookstores and coffeehouses that fall, the activists posted notices for a rally at the State Capitol. The posters featured a drawing of a wide-eyed bunny, dead center in a target. The copy read, "They're coming for the sacred sites, they're coming for the trees, they're coming for the homes—you're next." In its paws, the bunny held a monkey wrench, a symbol of the Earth First! movement.

The morning of the rally, a crowd of about 150 stood on the Capitol steps in the autumn sunshine, holding handmade signs saying "Stop 55! Save Sacred Sites!" Off to one side, Bob Greenberg pressed a cell phone to his ear. Suddenly he grabbed a bullhorn: he'd gotten word that thirty or forty police cars had blocked off Riverview Road. MnDOT was raiding the camp, he yelled, and was putting out the sacred fire.

The crowd disintegrated as everyone rushed back to camp. There, they found workers about to dig up the water, sewer, and gas lines to the illegally occupied houses. Sixty police officers and state troopers surrounded the workers.

Blocked by the solid line of officers, the people from camp watched helplessly as bulldozers lumbered down the street. Jim and the AIM activists trailed in the big machines' wake, beating a drum and singing. Women chanted in high, keening harmony, "We all COME from the MOTHer earth, and to the earth we SHALL return." The lament for Mother Earth rose in odd juxtaposition to the setting: the lawns, driveways, sidewalks, and street lamps of Riverview Road.

On the bullhorn, Bob Greenberg warned the protesters to prepare for tear gas. Kerchiefs over their faces, around a dozen activists sat down in front of a police van, linked arms, and prepared to be arrested. But when workers finished, the police backed the van down the street and disappeared. They had made six arrests, not counting one "catch and release," as one officer put it. The activists were left sitting awkwardly in the street. Afterward, a Department of Transportation spokesperson acknowledged that they had timed the shutoff to coincide with the rally at the Capitol. "Carpe diem," he said. Seize the day.

That fall, Carol had just a few months left in her home. She had been given until early December to move out. The Department of Transportation had offered her good money for her modest house. The latest offer had been $125,000, almost

Police officers and state troopers escort bulldozers sent to dig up water, sewer, and gas lines to illegally occupied houses in the path of the highway. Photograph copyright 2001 *Star Tribune*/Minneapolis–St. Paul.

twice the appraised value. "'Oh, Mom, you've got to take it. You could lose everything,'" Carol remembered her two grown children urging. But she stalled for months. "I wasn't ready yet. That was the hardest thing I ever did in my life, was to sign my house over to MnDOT," she said.

But she signed the papers at her kitchen table that summer. She went outside, distraught, to sit in the yard. "I just felt I was letting down all those people who were living around me. I knew once I was out of there, they weren't safe anymore," she said. But Jim came up and hugged her. "Jimmy said, 'Carol, you've got to let it go. You held them off until we got here, and now it's up to us.'"

Carol moved out of her house December 9. Moving men shouldered furniture out of the tiny living room and down the narrow staircase from the one-room attic, with its windows overlooking the trees and the river. When it was time to get Al into the car for the ride to the new house, he hung back. "He yelled, 'I'm not a goddamned dummy. I know I gotta go,'" Carol said.

When the house was empty, Carol's daughter, Deb, cleaned it from top to bottom and rearranged the curtains. "Goodbye old house . . . I love you little house," Deb wrote on the walls of the big attic room. Just before they turned the keys over to MnDOT, Carol and her children visited one last time. They climbed the stairs to the empty attic and looked out the windows. In tears, Carol added to what Deb had written on the walls: "Goodbye little house." Finally, Carol relinquished the keys—all except one, which she gave to Jim.

With Carol gone, people in the camp expected a police raid any day. The camp's phone-calling network stood ready to alert hundreds of supporters at a moment's notice. One person would telephone five, who would each call five more, and so on down the line. The network could spread information with lightning speed—and false alarms just as quickly. It was like the children's game of Telephone, in which a message changes as it spreads.

On December 10, the day after Carol moved out, the sun shone brightly and there was still no snow on the ground. The activists were holding a party to celebrate the four-month anniversary of the Minnehaha Free State—and also to pack the place full of supporters in case the raid came that day. People played fiddle and guitar and danced under the big oaks, dreadlocks flying. Many were barefoot; the day was strangely warm for early winter in Minnesota.

On one of the roofs, the activists had built wooden scaffolding draped with a hand-painted banner that showed a figure standing in front of three tanks. Underneath was the caption "The People Will Prevail." A cargo net hung in between two of the houses: a huge hammock for people to lie in when the police came. Earth First!ers looked down from the scaffolding, basking in the sunshine.

Someone had made a long banner and laid it out on the sidewalk in front of the houses. The words all ran together; to read it, you had to walk down its length. "CONGRESSSHALLMAKE-NOLAWREGARDINGESTABLISHMENTOFRELIGIONORPROHIBITING-THEFREEEXERCISEOFRELIGION" it said.

Throughout the party, the everyday details of camp life continued undisturbed. People wandered out of the garage kitchen with plates of food. A phone rang in a back room of one of the houses. A mattress lay on the floor; a blanket hung in the doorway. On the ground outside, extension cords led here and there. A young man coming out of the garage declined to say whether the houses had electricity. He replied pleasantly that they had phone lines but "how electricity is gotten and not gotten—these are problematic questions." It would be better to find out by asking someone you know, he said.

A young woman standing on the sidewalk with a clipboard signed people up for security shifts. Some fifteen to thirty people slept at the camp each night, she said. Two people called Bear and Sparrow were on the watch list so far.

The utility people had come by and snipped the wires so the vacant lots at the end of the street were dark at night, she explained—a good staging area for a raid.

CHAPTER 2

Operation Coldsnap

On St. Paul's Summit Avenue, nothing seemed out of the ordinary that Friday, December 18. As usual, there were few signs of life outside the street's historic mansions, which sat impassive behind their walls and hedges. The governor's mansion, with its carved stone balcony, slate roof, and many chimneys, was also quiet. There was one hint of activity, though: the parking lot tucked in back was full.

Inside, a basement room was so packed with people that the air was hot and close, even on a freezing winter day. Outgoing governor Arne Carlson, Minneapolis mayor Sharon Sayles Belton, and Hennepin County commissioner Peter McLaughlin sat on couches and overstuffed chairs, alongside officers from the State Patrol and police, fire, and sheriff's departments. Folding chairs held the overflow. The state legislators whose districts encompassed Riverview Road were also there, as were MnDOT officials, including the agency's chief spokesperson, Bob McFarlin. Former pro wrestler Jesse Ventura, who had shocked the nation by becoming Minnesota's new governor-elect just a few weeks earlier, sat quietly in the background. The mansion was decorated for Christmas, with evergreen garlands and stuffed toys to be given away to children, but the mood in the basement was not festive.

Officials had taken great care to keep the meeting secret. Someone had prepared what is known as an "if asked line" in case reporters wondered why so many top officials were suddenly unavailable. The officials were to say only that the governor was "getting a briefing" on the Highway 55 situation.

Tactical specialists from the State Patrol and the Minneapolis police stood before a poster-sized photo tacked to an easel. It was an aerial view of Riverview Road: the roofs of the seven houses, their yards, and the riverbank behind. With a pointer, the officers explained the plan of attack and the perimeter that would need to be secured. They were taking every precaution. Their undercover agents had infiltrated the protest camp and reported disturbing news.

The agents said that one of the illegally occupied houses on Riverview Road was locked and only Earth First! leaders had the key. They suspected there were explosives, weapons, or chemicals in the locked building. They reported rising tensions between Native Americans and Earth First!ers in the camp, a situation the agents feared could make the situation even more volatile.

As if that were not enough, they said, someone had tapped into the gas line that supplied Carol Kratz's house and had run a pipe to the house next door. The officers worried that a jury-rigged gas connection would cause an explosion or a fire, hurting or even killing protesters or bystanders.

And finally, they had heard rumors of booby traps in the squatters' houses. "The floors in there, they looked solid," said Don Davis, then Minnesota's commissioner of public safety, telling the story later, "but you'd fall through, just like you do in the jungle."

Davis, a tough old cop with thirty-one years on the force,

had no patience with the protesters or their cause, which he defined as "Let's save the trees and screw government." He used words like *malarky* and *shenanigans* to describe the protest, but in his view it went beyond that. "Some of them admitted they were anarchists," he said later. "It was their own words."

He had driven by the protest camp one day. "A cheap, run-down carnival is what it looked like to me," he said. He could understand why the neighbors across the street didn't want to "look out at that crap every day."

If Davis had gotten his way back when it all started, he was certain the situation never would have gotten so out of hand. The protest wouldn't have lasted more than a few days, let alone into December. "I'd have shut that thing down in August," he said. But back then, Davis had been talked out of doing anything. There was a problem: the elderly couple—what was their name? Krantz? Kratz?—who lived in the midst of the protesters. Davis did recall that the couple didn't mind having the protesters around, and even seemed to sympathize with them. His private theory was that the couple suffered from what's known as the Stockholm syndrome, named after a famous case in Sweden in which hostages came to sympathize with their captors. Whatever the case, as long as the old people lived there, it was difficult to dislodge the protesters.

Officials at MnDOT hadn't wanted to seem like "uncaring ogres," Davis explained. They could imagine the television cameras rolling as two kindly old people were thrown out of their home. The agency's spokesperson, Bob McFarlin, had argued that it was best just to wait.

To reporters covering the Highway 55 story, McFarlin was a pleasant, knowledgeable voice on the phone, a man who always returned calls promptly, but he was really more than just a spokesperson. He was also transportation commissioner James

Denn's chief of staff. McFarlin understood how the world of power worked. And he was concerned about MnDOT's image.

After all, he figured, the city and the county, not MnDOT, were the ones pushing to get Highway 55 completed. "They wanted the project badly," McFarlin said, telling the story later. It would smooth the way for Minnesota's first light rail transit line, which was to be built between downtown Minneapolis and the airport and the nearby Mall of America.

If the city and the county wanted light rail, McFarlin reasoned, they should help get rid of the protesters. Until then, MnDOT was prepared to wait. "For the image of the agency, I wanted to make sure we weren't the sole villain," McFarlin said. During MnDOT's deliberations with law enforcement officials in August, McFarlin's arguments had prevailed.

But the pressure to oust the protesters had grown since then, like a vague pain that builds to a chronic headache. The legal residents of Riverview Road had become more and more upset. "A Protest movement that started rather quietly two months ago is getting worse and worse," a man who had lived on Riverview Road for sixty-two years wrote in letters to his city council representative, the mayor, the commissioner of transportation, and the governor. "Let me say we fear for our safety.... We implore you to listen to us ... to clean up this sorry situation and return our neighborhood to some degree of normalcy."

"You are sending the message: Do what you want.... The City will do nothing!" another enraged resident wrote to the city of Minneapolis. "I feel sorry for the next neighborhood that is invaded."

But McFarlin refused to budge without the city and county. "The point was, *MnDOT* wasn't going to do anything—we were *all* going to do something," he said.

So the months went by, and the protesters settled into

Riverview Road as if they owned it. "They had patrols out to patrol the police, and would challenge the police and their right to be there, which was, I thought, an interesting approach," Minneapolis police captain Bud Emerson said dryly. Emerson has a quiet sense of humor and a knack for understatement. At the time, he was day watch lieutenant for the south Minneapolis precinct that included Riverview Road. It was his thankless job to ride herd on the protesters, but he recalled them good-naturedly. "We had this little dance," he said of the two sides.

His officers would post "No Trespassing" signs, and by the time they had reached one end of the block, the protesters would be taking the signs down on the other end. Emerson would go out to the camp to talk to the "nominal leadership" of the group and suddenly they wouldn't be there; they vanished into the woods like Robin Hood and his merry men.

When the horse corral appeared, officers had to go out and tell the protesters it was against the law to build a corral in the city. Although in that case the horses disappeared a few days later, what the police said seldom made much difference. As Emerson recalled, "We'd go out and say, 'Here's what we're hearing,' and they'd just smile at us and say, 'Oh, really?'"

Emerson figured it wasn't worth wasting much time chasing protesters through the bushes. Trespassing was not at the top of the "pantheon of crimes," as he put it. And Emerson realized how dedicated many of the protesters were. He understood why they cared about the wild places around Riverview Road. He lived just two blocks away and loved the woods himself. One of his favorite winter pastimes was cross-country skiing through the urban forest that still stood in his neighborhood. Sometimes the protesters would ask him how he could stand by and watch as the road went through it. He told them, "If I were king, I might have wished to do things differently, but I'm not a king, I'm a lieutenant."

Emerson encouraged the protesters to go through the courts or the state legislature, but they said they didn't have enough money for the courts, or enough community support to get help from legislators. He told them, "There's a message there. There are minority rights but they don't get to run the show. I think you're stuck."

He was surprised sometimes at what poor strategists the activists were. "The protesters always thought everybody going through there was a spy," Emerson said. Ordinary citizens used to watch the activists from the bushes through binoculars, as if they were a circus or reality TV. Then, behaving like outraged landowners, the protesters would escort the voyeurs out of the area. They seemed oblivious to the possibility that they were alienating a neighborhood they had hoped would back them.

When complaints came in about urinating in the bushes, Emerson explained to the protesters that if they wanted to build support for their cause, they would have to stop doing that. After the water lines to the illegally occupied houses were cut, some protesters maintained that they had to use the yards as bathrooms. If Emerson pointed out that they were trespassing, they said they were following a higher law. So he told them they could still flush the toilets by pouring buckets of water down them. "They all kind of looked at me like, 'Oh yeah, man, cool,'" he said.

Emerson himself never saw the jury-rigged gas line that the undercover agents reported, but he once got a message at the station saying that one faction had hooked up the line over the objections of another. Whenever inspectors arrived, protesters swarmed around them. By the time they got to where the connection would have gone, the officers figured someone would have had a chance to get rid of the piping. One officer saw an empty trench where a pipe might have been. And the police knew

Jim Anderson had been a pipe fitter. "He apparently knew what do," as safety commissioner Don Davis put it.

Later, some people who had been in the camp flatly denied any hookup. Others said a small group had done it over the objections of the rest but disconnected it soon afterward. But by that cold day in December when the officials gathered at the governor's mansion, whether the hookup actually existed or not no longer really mattered. Something terrible had happened just one week earlier; now even the suspicion of a jury-rigged gas connection was enough to provoke the full force of the law.

On December 11, 1998, a gas line exploded in St. Cloud, Minnesota, killing four people and injuring fifteen. Witnesses described steel beams flying into the street and bricks and boards blown blocks away. Don Davis had looked down at the rubble from a helicopter a few hours after the blast. "It was a terrible sight that day, a huge hole in the ground. It was like a bomb had been dropped," he said. Governor Carlson was in a second helicopter. Later the governor asked Davis if the same thing could happen at the protest camp, and Davis told him it could. "He said, 'That does it,'" Davis recalled later. "'Do what you have to do.'"

Hennepin County commissioner Peter McLaughlin drew the same conclusion. For weeks, he had been hearing rumors of the gas line connection at the camp, which sat in his district. Now the St. Cloud explosion was a shocking illustration of the dangers of a gas line leak. "I did not want to have it on my record or on my conscience if an explosion occurred down there and either the protesters or the neighbors were hurt or killed," he said, telling the story later. "That's a reality and I'm elected to make judgment calls like that. The danger was too great. I couldn't live with it if the thing blew."

A Democrat in the Hubert Humphrey tradition, McLaughlin represented some of the poorest neighborhoods in Minneapolis; he worked tirelessly for civil rights, affordable housing, and other liberal causes. But McLaughlin had no sympathy for the protesters. Long afterward, he still got angry when he was asked about them. That they had portrayed themselves as environmental heroes particularly irked him; he felt it showed they had no sense of history, no inkling that the battle over Highway 55 had been raging for decades.

The real heroes, McLaughlin believed, were the ordinary citizens who had fought the highway years before the protesters set up their tents, people who sat through thousands of hours of meetings to stop the project as it was then envisioned. "They took on MnDOT when it was a behemoth that had never been stopped," McLaughlin said, his voice rising. In those days, the 1950s and 1960s, MnDOT, then known simply as the Highway Department, was leveling block after city block to make way for interstate highways. In Minneapolis and St. Paul, the agency was building multilane freeways sunk in deep trenches that slashed through the cities.

The original plan for rebuilding Highway 55 called for just such a project. It seemed unbeatable. "*Fritz Mondale* signed the condemnation papers for a six-lane freeway," McLaughlin said, referring to a revered figure in Minnesota politics, Walter Mondale, then the state's attorney general. The highway department bulldozed hundreds of houses and small businesses on either side of a stretch of Highway 55, leveling space for the six lanes plus four lanes of frontage road.

"They were God Almighty, and they could do whatever they wanted to," a local citizen named Walter Bratt remembered. "In those days, freeways went through." Bratt and his wife, Carola, had seen Interstate 94 rammed through downtown Minneapolis, ensnarling the Basilica of St. Mary and the grand old houses on Lowry Hill in a web of on-ramps and overpasses. They had seen the concrete trench of 35W slash through tree-shaded city neighborhoods, isolating the parts from each other and opening a wound that spread urban decay like an infection. As the river of concrete rolled steadily on, the Bratts and many other ordinary citizens were starting to have second thoughts. "It took about twenty years to rile up the people not only in this city, but in every city across the country," Walter Bratt said.

The Bratts first saw the plan for the new freeway that was to replace Highway 55 at a neighborhood meeting given by their city council member, who had been out in Washington lobbying for funding. When he asked if there were any questions, Carol spoke up. "Who needs another freeway in the city?" she asked.

"He was just stunned. He'd never heard a question like that," Walter Bratt said. Other politicians were starting to hear the same message. Finally, Minneapolis mayor Al Hofstede convened a citizens' committee to look at the Highway 55 project. Walter Bratt was appointed to the task force. An engineer with a background in geology, he told MnDOT he wanted to see a copy of the freeway design.

A few weeks later, a truck pulled up to the Bratts' house and delivered a three-hundred-page document. In that era before the photocopier, the house smelled for weeks from the ammonia used in the copying. "I think I'm the only one on the task force who read all this stuff," Bratt said. He studied the layout and the geologists' reports. He paid particular attention to a tunnel that would run underneath Minnehaha Park. He noticed that it would cut into the rock underlying the park's beloved landmark, Minnehaha Falls.

It was a "monstrous project," Bratt told MnDOT. He used his expertise as a geologist and the agency's own data to show that the tunnel would threaten the hydrology of the waterfall. "You're just setting yourself up for a huge debacle here, destroying the whole falls," he remembered saying. From that time on, he recalled, the tunnel idea was as good as dead.

Next, Bratt and his allies on the committee drew up their own plan: a four-lane, forty-mile-per-hour, ground-level high-way—not a trenched freeway. And they tried to sell MnDOT on the then-novel idea of using light rail transit to absorb part of the projected traffic. Bratt was an early and ardent proponent of light rail. He had read of the first efforts to build light rail systems around the country. A friend of Bratt's who was a model-railroad buff built a miniature version of the old Minnehaha streetcar line and took it around to meetings Bratt helped arrange. "People were really fascinated by that little model," he said. Bratt's hand-drawn plan for the Hiawatha light rail transit line showed each stop, from the airport to downtown Minneapolis.

But the Highway Department's response was unenthusiastic; Bratt suspected that the task force's real purpose all along had been to sell citizens on the freeway. The mayor and the city council convened a second task force, which held dozens of public meetings. Its members considered an eye-glazing number of alternatives, but in the end the plan they chose was remarkably similar to Bratt's original. Both contained a detail that nobody at the time paid much attention to—rerouting a short stretch of the highway along the Mississippi, through Riverview Road.

To Bratt, the reroute made perfect sense. The swing along the river would run the new stretch of highway right along an old railroad right-of-way. It would take only a few houses, instead of many more that would have to be bulldozed if the existing Highway 55 were widened in place. It would make constructing a light rail line much easier, Bratt figured. He didn't give it a second thought.

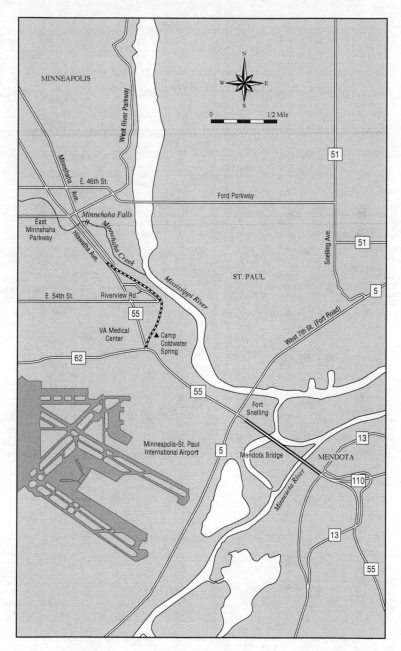

Area map of reroute of Highway 55.

Years later, after the protesters had taken their stand by the river, Bratt got a map of the planned reroute and trudged the length of it, taking notes. It was winter and sixteen degrees below zero—bitterly cold even for Minnesota. He was sixty-nine years old. Bratt looked at all the trees that would be taken by the highway. As far as he could see, just about the only "good trees" were the four oaks. A lot of what had to be cut were weed trees, he thought, box elders and poplars and such. Considering that you have to have roads in the city, the plan was the best that could be expected, he figured. The protesters must just want something to protest about.

Hennepin County commissioner Peter McLaughlin agreed that it was the best possible plan—not perfect, perhaps, but as close to it as the political process allowed. "That was the deal," he said flatly. "An agreement was reached." McLaughlin conceded that values had changed since the deal was sealed in 1985. ("In the eighties, we cared about housing," he said bitterly. "Now we care about trees that are supposedly sacred.") But it was still an agreement, negotiated by citizens who had worked hard. McLaughlin believed in the just workings of a system in which deals are struck, hands shaken, words kept. It frustrated him that the protesters didn't seem to understand that years ago the community had won a great victory over a mammoth freeway. Now, after a thirty-nine-year process, here were these eleventh-hour protesters, acting as though they knew better than the citizens who had worked so long. Besides, McLaughlin insisted, these things could not be reopened again and again. It was time to move on.

At the mansion, Governor Carlson listened carefully as the tactical experts explained the plans for what the police called Operation Coldsnap. Since the element of surprise was everything, the raid would come at night. The tacticians hoped for bitter cold, to lessen the possibility of large, unruly crowds. Every aspect of the operation had to be carefully controlled. The Minneapolis police and fire departments, the State Patrol, and the Hennepin County Sheriff's Department would all play a role. A huge number of officers would be called in from across the state. Even public safety commissioner Don Davis was startled when he heard the figure. "I almost fell off a chair," he said later.

Davis is not a person with a weak stomach for police action. A few summers earlier, he had cheered the governor's decision to send State Patrol helicopters sweeping across the night skies of Minneapolis in response to a spike in the city's murder rate. When neighborhoods woke at three o'clock in the morning to the chop of helicopter blades and people squinted at the searchlights raking their backyards, Davis had no qualms. "Little gang bangers came out like bugs from under a rock," he said; the murder rate dropped, and that was what mattered to Davis. Now, although the number of officers required was even higher than the figure he had heard just a few days before, he reasoned that the tacticians knew what they were doing. The overall objective was safety, after all.

And there was one more disturbing consideration. Small children had been spotted out at the camp. Davis had briefed Governor Carlson, and the governor was concerned. After all, kids didn't have a choice when their parents put them into a potentially dangerous environment, like a camp where there might be an explosion.

Later, Davis didn't remember anyone using the word *Waco*, either before or during the governor's mansion meeting, but MnDOT's Bob McFarlin said that Governor Carlson worried aloud that in his last few days of office, a second Waco, this time in Minneapolis, would forever tarnish his legacy.

The Republican governor looked at the Minneapolis politicians, many of them Democrats. They had criticized and second-guessed him the summer when he had sent the helicopters over Minneapolis, even though things were so bad that the *New York Times* was calling the city Murderapolis. Now, according to McFarlin, the governor was "terribly paranoid," worried they would blame him if anything went wrong. "He kept saying, 'If we have another Waco on our hands here, I know what you're going to do. You're all going to point your fingers at me.'"

McFarlin and others who were at the meeting agree on how it ended: Carlson put the question to all the high-level officials gathered at the mansion. "He was in the middle of the room and he pointed all around," McFarlin recalled. "He said, 'I want to hear you say it if you support this.'"

The vote was unanimous. The governor gave his approval.

CHAPTER 3

The Raid

It was past midnight, and Jim Anderson struggled to stay awake at the steering wheel. He had been patrolling the streets around Riverview Road for four nights now, watching for a raid.

For weeks, rumors had swept the camp: the National Guard would be mobilized; someone who had overheard soldiers talking in a coffee shop was dead certain. Someone had gotten a tip from somebody whose brother-in-law worked for the State Patrol, and the raid would come that weekend for sure. At each false alarm, people would lock down, chaining their arms to the dragons in the houses. They slept when they could and were more exhausted each time.

Now Jim's cell phone bleeped; a lookout reported a cluster of patrol cars headed for the airport. Jim sped to a hill a quarter mile from camp with a wide view of the airport's runways and acres of well-lighted concrete. The line of patrol cars and heavy equipment seemed to stretch for miles. He had never seen anything like it. It looked like a war movie. *Holy cow*, he thought, *did we get them mad now*.

"They're milling up," he radioed back to camp. "It's going to happen tonight."

"How many cops?" someone asked on the radio.

"A lot," he said.

Jim drove swiftly back to Riverview Road, taking an over-pass across Highway 62, a main artery into south Minneapolis. The highway was empty. The police had it blocked off.

Earth First!er Bill Busse was keeping watch outside North House when he saw the first headlights. A single pair turned off the highway onto Riverview Road. As it took the corner at a lighted gas station, he could see the yellow side of a moving truck, and the word RYDER in black letters. The station's lights caught other dark shapes in quick succession: a line of trucks with their headlights off. Within seconds, the trucks were jolt-ing into the camp, one truck for each of the seven houses, hit-ting the curb hard. The doors of the trucks burst open, and SWAT teams armed with submachine guns and semiautomatic pistols jumped out. Busse yelled out a warning. He heard the wail of the camp's siren and officers yelling, "Get down on the ground, get down on the ground!" He pulled out his cell phone and began making frantic calls as shapes seemed to come at him in the dark from all directions and the street filled with headlights.

One by one, the squad cars streaming north from the air-port turned onto Riverview Road and parked bumper to bumper. Busloads of officers pulled into the woods by the Mississippi. Within minutes, they sealed off the camp with yellow tape: Police line. Do not cross.

A helicopter circled overhead, the thump of its blades a steady beat above all the other sounds: voices yelling, engines idling, dogs frantically barking in backyards. Hook and ladder trucks bumped across the lawns, flashing red. On the roof of the house with the Earth First! banner, a young man in a Santa Claus suit stood chained to the chimney. Caught in a spotlight, he looked bizarrely jovial, like a mechanical department store

Santa transported to an accident scene. The lights blazed on Joe Hill locked into his tripod beside one of the houses, casting long shadows. Over everything there was a weird calm as the first snowfall of the season drifted gently down.

The camp's phone network had kicked into high gear, and soon a crowd of about a hundred supporters milled behind the police line, unable to cross it without being arrested. The temperature was ten degrees above zero, as dauntingly cold as the raid's planners had hoped. At the edges of the spotlights, the crowd could see only moving shapes and shadows. It was impossible to know what was happening to their friends in the dark basements. "Don't hurt them!" some screamed as police carried out limp, handcuffed forms. "It's Meaghan!" somebody yelled. "Hang in there, Meaghan! We love you, Meaghan!"

When the raid began, two young activists called Natalia and Dr. Toxic were waiting in the basement of Kitchen House, hidden from the outside world. On the two-way radio they held between them, everyone tried to talk at once. "How many cops?" they heard someone ask. "Oh my God," a voice said. "Too many to count."

Except for the embers glowing in the fireplace, the room was completely dark. A double dragon, one with two armholes, was sunk in the basement floor; the excavated dirt was piled against a barricaded door. The basement's only entrance was now a hole cut in the bottom of a first-floor closet, just wide enough for a slim pair of hips. It was the best-barricaded spot in the camp, Natalia thought, and it would take the police a long time to break through.

Natalia had hardly left the basement for days. She wanted to be there when the police came. Barely eighteen, with big gray-blue eyes, porcelain skin, and glossy auburn hair that she buzzed short, she had already been arrested in animal rights demonstrations so many times she'd quit counting. She had dreamed of being an activist ever since she was a little girl living in a Washington, D.C., suburb. She wanted to be like her grandmother, an "amazing lady" who had gone to Central America during the 1980s and who displayed her arrest record proudly on the refrigerator. "I thought activists were awesome," Natalia said later. "And that was what I wanted to do. I wanted to make a difference in the world."

Dr. Toxic was there only because Natalia had asked him to be. A post in the basement was the last thing he would have chosen for himself. A lanky twenty-one-year-old with a partial crew cut and a long ponytail, he had been arrested half a dozen times as an animal rights activist, and he knew there was something about him that police didn't like. He had seen how officers gave him a certain look and then tightened up his handcuffs a notch so that they really hurt. Earlier, another person at camp had asked him to take a basement lockdown post and he had declined. "I said no, I don't want to be in a basement 'cause I'd get my ass kicked," he said later. "Behind closed doors they can, and do, do anything they want to. I hate to put myself in that position. More times than not they are going to do what you dread." But he agreed when Natalia asked, because they had just started dating.

Natalia had called him during several of the many false alarms. She was brave and feisty, but at only five feet tall and barely a hundred pounds, she wanted someone she could trust by her side when the police came. Earlier that night, Natalia had telephoned Dr. Toxic at a Christmas party at his family's home in Osseo, a Minneapolis suburb. He left at midnight, not telling his parents where he was going. On the drive in, he heard Jim on the radio saying that this time the raid was really happening.

Dr. Toxic slipped through the hole in the closet floor and

climbed down a rickety homemade ladder. He and Natalia held each other briefly in the freezing dark, then, lying side by side, each plunged one arm up to the shoulder in the double dragon.

Later, Natalia and Dr. Toxic remembered what came next only as sounds. They heard screams and glass breaking, the camp's siren, heavy footsteps, and police yelling, "Get down! Get the fuck down!" Dr. Toxic had been in the army; he recognized the sound of gear and weapons jolting as the men ran. He told Natalia to put her free hand on top of her head so the police could see it.

On their handheld radio Dr. Toxic and Natalia heard people coughing, then high electronic bleeping; the police had jammed the radio frequency. The two had just enough time to put on gas masks before they heard what they mistook for an explosion; it was the State Patrol battering down the barricade. In seconds, troopers were in the room, the beams of their headlamps sweeping through the blackness.

"We entered with guns drawn," one of the troopers later testified at a court hearing. "We weren't sure if they were armed or not. We weren't sure what kind of resistance we would meet." But only silence greeted them. "There was an old couch, a fireplace, some junk laying around, quite a bit of junk laying around, and there was two people laying face down on the floor," the trooper testified. "I couldn't really tell you what they looked like other than the gas masks at that point."

The officers commanded the figures on the floor to put their hands above their heads, but they didn't move. One trooper tried to pull Natalia's locked-down arm over her head, then realized the arm was held fast beneath her. Other officers tried to get Dr. Toxic to release his arm. When that failed, they used what's known as "pain compliance," a standard police procedure. It means applying pressure in ways that hurt but don't cause permanent damage. One officer bore down on Dr. Toxic's head and neck. Two others worked his calves. "I attempted to reason with the male," one trooper wrote in the arrest report. "I explained to him that I was aware of how the lockdown device worked, and I was aware he could remove himself at any time very easily. I literally pleaded with the male to stop resisting and told him I would be forced to use additional force."

Dr. Toxic remembered cops yelling at him to unlock, lifting both ends of his body, bending back his head, and choking him. Then he felt pepper spray: a hot stinging pain across his face. His eyes teared; mucus streamed from his nose.

"I sprayed the product into—I had gloves on my hands, leather gloves, and I sprayed the product on my glove hand and I wiped it across the face," one officer testified later. "The entire facial area, from forehead to chin, ear to ear."

Moments passed, but "the subject was not yet complying," another officer wrote in his report. "I then used my flashlight (as an impact weapon) to strike the subject on his right calf (tibia) . . . the subject was still not complying. . . . I then struck the subject a second time."

Finally, Dr. Toxic unlocked. A trooper handcuffed him and led him upstairs. "He was resisting passively, using dead weight, wouldn't walk at times. Sometimes he would and sometimes he wouldn't," the officer testified.

"Look at your little boyfriend there, isn't he tough?" Natalia remembered one officer telling her as Dr. Toxic was led away. She had been talking back to the police, pretending she didn't know what a lockdown was, and "being a smartass," as she put it. She refused to unlock when they tried the pain compliance holds. She tried to stay calm, to curl herself into a small, quiet space. Then, she remembered, they said, "We have something for you" and showed her a yellow and red package.

"I had monitored Sgt. Freng as he spoke with the female,"

the police report reads. "I inquired with Sgt. Freng if I should apply the chemical to the female's facial area. He indicated I should. I applied the chemical in the same manner as the male. Within sixty seconds, the female indicated she would remove her hand from her lock down." A different officer's report gives an additional detail: "The female subject still would not comply so I applied pressure to the mandibular angle and she unlocked herself also." To Natalia, it felt as though the pressure on her jaw came from something cold and metallic. "OK, listen, bitch," she remembered a voice saying. "Unlock or I'll shoot."

Later she would testify in court that an officer put a gun to her head. The officers' reports make no mention of it, though, and with no outside witness, it was her word against theirs.

In their basement lockdown two houses away, two more activists, Solstice and Marshall Law, heard a sudden loud hissing and realized that pepper gas was seeping through the barricaded walls. Their plan had been to talk to the media on their cell phones as the raid unfolded, but now the gas was choking them and the raid had barely begun. Through the walls they heard officers yelling at them to come out.

Solstice, twenty, had been an activist since high school. He had encountered mean cops and cops who weren't comfortable with what they were doing, but they had never sounded like this. He realized suddenly what it was: these cops were terrified. He could hear it in their voices. They were "way scared," he said later.

Solstice fumbled for his cell phone and dialed a reporter friend in New York, but all he got was an answering machine, and anyhow he could barely talk, he was coughing so much. The basement wall began to shake under the blows of a battering ram. "They started screaming to us, 'When this wall comes down, you'd better have your hands up, because we have our guns drawn,'" Solstice said later. His plan to talk live with the media evaporated; his mind wasn't working clearly. He couldn't think who to call next, and it didn't seem worth it anyway. "We really didn't want to get killed over this one," he said later, "which honestly I'd never felt was such a possibility." The basement wall shook harder. It was starting to give way.

Solstice and Marshall dropped their cell phones, afraid that anything in their hands might be mistaken for weapons. They released their other hands from their lockdown and ran over to the wall. "We're trying to come out," they yelled. Solstice picked up a crowbar and ran to another door, terrified that if the police broke through and saw him with it they would shoot him. Frantically, he pried the door open. Beyond it was an opening to another room. They crawled through, held their palms in front of them to show they had no weapons, then clambered out.

"I saw a room full of, it looked like eight to ten cops, all in gas masks, with laserscope rifles pointed right at us," Solstice said later. As soon as he and Marshall were in custody, "there was just an air of relief like I've never experienced over the whole room."

The SWAT teams who burst into the houses had reason to be afraid. As one officer wrote in his report, "The intelligence that our team received included information that individuals at the residence had a wide variety of weapons such as clubs, axes, and edged weapons and that their mental state was as such as these weapons may be used against us."

Minneapolis police inspector David Indrehus, one of the raid's main planners, knew how terrifying such situations could be. "You know, if you think about yourself sitting in your house, God forbid some night and the door comes flying off the hinges and in comes this entry team with guns and everybody screaming, *Down, get down*," he said later. "If you think that's a frightening, horrible thing for people on the receiving end, I'm telling you it's not much better for those doing it."

But when the teams stormed into the houses on Riverview Road, they found none of the dangers they had been warned against. There were no weapons, chemicals, or explosives. There were no booby traps. If, as safety commissioner Don Davis thought, it had been possible to fall through the floors as if you were in a jungle, no one did.

Solstice and Marshall Law were astounded when the officers asked them where the booby traps and weapons were. "We were just, like, are you crazy? This has been a nonviolent campaign for four months now. Do you think we would be that stupid and suicidal?" Solstice said. "And when we said we were nonviolent and explained that to them, they would laugh and mock, as if they had inside information, like they knew we were full of shit."

Safety commissioner Don Davis rode into the camp just behind the Ryder trucks, in the jumbo-sized mobile home that served as a command vehicle for Operation Coldsnap. He would have preferred to ride in one of the Ryder trucks, but he knew he would just be in the way. Going in second was the next best thing. He stepped down from the mammoth vehicle and surveyed the scene. Later, he retold the story with relish. "In case you haven't noticed," he said, "I love being a cop."

His most vivid memory was of some guy in a Santa suit, shivering on a rooftop. "What do you want to do with him, Commissioner?" Davis recalled an officer asking him. "I said, 'Leave him.'" The guy got plenty cold up there, Davis said, and by that time the other protesters had been removed. The operation was proceeding flawlessly, Davis thought, just according to plan.

In the woods and yards, the officers lit bonfires and stood silhouetted against the flames. The trees along the river loomed

An activist (Wes) dressed as Santa Claus huddles on the chimney of one of the illegally occupied houses during the police raid in December 1998. Photograph copyright 2001 *Star Tribune*/Minneapolis–St. Paul.

in the background, firelight and shadows playing across their trunks. The sweat lodge, tipis, and tents were gone, along with the "Minnehaha Liberated Zone" signs.

By the time Jim Anderson got back to the camp, it had been sealed off. Behind the yellow police tape, the ceremonial fire he had helped tend for four months was ashes. "They come here five days before Christmas and they put out our sacred fire—that's like ruining an altar at a church," he told a reporter. "They went back here in our *M'dewin* sweat lodge, they tore it down and *burned* it," he said incredulously. "It was on fire right back there, we witnessed this, I've got it on tape." His words streamed out, one thought tumbling into another. "Here comes Arne Carlson now," he said, hardly pausing for breath. "He went over there and warmed his damn hands on our sweat lodge. Arne Carlson's a rat fink."

As the operation wound down, Governor Carlson strolled through the crowd. Politician style, he was hatless in the bitter cold. He had spent a sleepless night, staying in close contact with

Officers stand silhouetted against the flames of bonfires lit during the raid of the protest camp on Riverview Road in December 1998. Photograph copyright 2001 *Star Tribune*/Minneapolis–St. Paul.

an aide to make sure everything went well, then arrived at the scene just before sunrise. "I'm the one responsible for the operation. The decision ultimately was mine," he told reporters. One of them asked why he decided this was the best tactic. "Well, what tactic would you recommend?" he asked sarcastically.

"As opposed to, say, waiting them out till spring," another reporter said.

"Oh, my heavens to Betsy, what a terrible decision!" replied the governor. "We have pipelines here, these people have illegally tapped into those gas lines. Those gas lines could have blown at any time. These houses, with innocent people in them, their lives would have been placed in jeopardy. I'll tell you who was harmed here. This lovely neighborhood, and I used to live here."

A reporter asked how close by, and Carlson couldn't remember exactly. He started to ask his aides, then interrupted himself. "Oh, you guys wouldn't know that," he said.

Wes Skoglund, the state representative whose district included Riverview Road, arrived just as the sun was coming up. For months, the street's legal occupants had been pleading with him to get the protesters removed, and he had assured them something would happen soon; he just couldn't say when. Now, as he approached the legally occupied side of Riverview Road, he experienced something that had never happened to him in all his years in politics. Men and women crowded around, reaching out their hands to touch him. "They were calling 'Wes! Wes!'" he said later. "They were so thrilled to see those guys gone."

The joint press conference at the Third Precinct station later that morning was what MnDOT's Bob McFarlin would remember as a "crowning achievement." At last, the city and the county were standing behind MnDOT. The elected officials who represented Riverview Road stood by as Minneapolis mayor

Sharon Sayles Belton stated that she had been fully informed of the operation and supported it. "It allows the public to move forward on a project that is very important," she said. "I want to just say personally that I believe that citizens do have the right to protest, but that protest should be responsible, and when it becomes irresponsible it is appropriate and necessary for elected officials to take action."

The operation had been the largest combined law enforcement action in the state's history, safety commissioner Davis announced, and nearly six hundred officers had participated. (Later, the paperwork would show the actual number to be more than eight hundred.) Thirty-seven people had been arrested, including three who were later discovered to be juveniles. Already, an as yet unspoken criticism hung in the air: the operation had been much bigger than necessary.

"There will always be questions, 'Did you need that many officers?' The answer emphatically is yes," Governor Carlson said in response, then added a new justification. "We wanted to make sure that there was a factor of intimidation." The protesters, he said, were "basically anarchists" who had "substantially terrorized" the neighborhood.

Hennepin County sheriff Pat McGowan announced that twenty-six protesters faced probable cause felony charges for obstruction of justice with force.

"What was the force?" a newspaper reporter asked.

"The force was resisting the arrest," McGowan said.

"How?" said the reporter.

"Anywhere from attempting to not cooperate with the officers, struggling with them . . ." His voice trailed off.

Minneapolis police chief Robert Olson came to the sheriff's assistance. "They had lockdowns all around the area so they could insert their arm or some other part of their body and connect it to the concrete set in the ground and inside the house," he said. "They had quite a plan."

But the charges actually filed against the protesters gave no indication that they had used force. No protester was charged with anything more serious than a misdemeanor, and in most cases the charges were dropped soon after the arrests. Of the thirty-four adult protesters arrested, only Natalia's and Dr. Toxic's cases ever went before a judge.

Looking back on the raid later, MnDOT's Bob McFarlin conceded that the threat from the protesters had perhaps been overestimated. "I will admit there was a bit of 'black helicopter' going on," he said. (In right-wing extremist legend, black helicopters are mysterious unmarked aircraft carrying United Nations storm troopers on covert missions to prepare for the New World Order takeover of the United States.) But McFarlin pointed out that the transportation department and the politicians had had no information of their own about the situation at the camp. They had relied on law enforcement's assessment.

The raid's defenders also argued that even though the protesters hadn't turned out to be violent, there had still been the danger of the jury-rigged gas line. At the press conference after the raid, safety commissioner Don Davis announced, "We have documentation that in the past few months illegal and unlawful, improper gas line connections were made to that area."

But when reporters asked for the evidence that the gas lines were hooked up, the officers floundered. "We had it that they were. The pipes and the place, and the holes, and, yeah," said Chief Olson. There was a short silence. "They were, in fact, bootlegging gas out of the house," he added. "There's piping all over the place." Pressed further, Olson said it was copper piping, coming out of the Kratz house and spreading to the other houses.

Asked much later about the illegal gas line connections, Minneapolis police inspector David Indrehus said he thought they had been found and disconnected. He couldn't recall that they had taken any photographs of the pipes, though. He had a million things to think about that night, overseeing an operation comparable in scale and complexity to having the president come to town. The planners had thought of every detail, even disconnecting the entire citywide traffic monitoring system that night so some lonely soul on the overnight shift at a television station wouldn't notice anything out of the ordinary. But proving that there had been an illegal gas line connection hadn't been on the list.

"In afterthought, maybe that should have been a prime concern, so we could have presented it. I don't know." Indrehus said. "For purposes of justification, certainly we should have taken those pictures." But he himself had had complete confidence in the undercover agents' reports of an illegal hookup. "We were comfortable it was there. Our people were not going to lie to us."

It wasn't until after he was arrested that Solstice, the young activist rousted from the basement by officers pointing laser-sight rifles, found out there had been undercover agents in the camp. Rumors started to float around the jail that two of the arrested activists, a man named Ben and a woman named Leslie, had an odd story to tell.

Ben and Leslie had been standing outside the houses when the raid began. Later, as one officer was questioning them, a second officer approached and asked who they were. The first officer said, "These are the people in the photos." The second officer said, "No they aren't. Arrest them."

As the activists puzzled it over later, it seemed clear that Ben and Leslie had been mistaken for informants. But who were the real informants? The activists' suspicions lighted on a couple who called themselves Cat and Dog.

Natalia remembered Dog as a baby-faced man in his late twenties or early thirties, with dark hair and thick black-framed glasses. He slept on the couch upstairs from her basement lock-down. Once, when she ran out of candles, he had offered to take her to the gas station on the corner to buy some. Let's talk, he had said. She declined, thinking, "Ooh, you're weird," she remembered. Others recalled Dog's dark hair and "soda bottle" glasses, but not much more. About Cat, they had only the haziest memories.

"They came around at night," Solstice said later. "They hung around in the kitchen and ate doughnuts and stuff. They weren't very noticeable. A lot of people didn't trust them, but a lot of people didn't trust anybody." Cat and Dog were "vague and nondescript," he said, and besides, he had seen them only in the dark. Later, he figured that was no accident.

As Solstice and his friends compared notes, Solstice realized why he and Marshall Law had heard the hissing of pepper gas so early in the raid; Dog or Cat must have set it off. Solstice and the others figured Dog and Cat must have been the source of the police fears about weapons and booby traps. Those suspicions hardened after the raid, when Cat and Dog disappeared and were never heard from again.

Police reports of the raid confirm that "undercover Minneapolis officers" set off gas before the raid, but give no clue to their identity. Asked about it much later, Inspector David Indrehus declined to reveal the number of agents or whether any went by the names of Cat and Dog. The activists' detective abilities did not impress him. "I don't believe for a minute to this day they knew who our people were," he said.

Whoever they were, the agents' briefings had been key to the decisions made by everyone who planned and approved the raid, up to and including the governor. Afterward, when it was clear there had been no weapons, explosives, chemicals, or booby traps, there was no question of holding the agents publicly accountable for their performance. Their identities were sealed. Apart from the police and the planners, the only ones who knew of the agents' existence were a few activists. And if a scraggly-looking bunch like that were the only ones who knew the ill-advised reasons for ordering such a massive raid, the mainsteam citizens of Minnesota weren't about to listen, even if they did have to foot the bill.

The raid cost nearly $380,000—more than ten thousand dollars for each arrest. The city of Minneapolis had to ask the state legislature for special funding to pay for it, but if the public wondered why it had taken more than a quarter million dollars and eight hundred officers to arrest thirty-seven protesters, nobody said much. Months later, the legislature rejected the request for reimbursement, in part because the raid was seen as overkill, but even then only the activists seemed truly shocked by its size.

"It was like a stone wall came down," Carol Kratz said. To her children's dismay when they found out, she had left Al home alone and gone to the camp that night the minute someone called her about the raid. Carol had watched aghast as the armed SWAT teams leaped out of the Ryder trucks. "It was unbelievable," she said. "I just stood there. All I could say is, 'What are they doing?'"

"We had no idea it would be a brutal as it was," Tumbleweed said. "A lot of people lost their innocence that day. Any sort of belief in the system, any sort of belief in fairness … any sort of faith we had in America was over."

The bedrock belief among activists that police are ruthless and brutal seemed to have been amply demonstrated. If Meaghan had ever had doubts about what she read in the alternative papers, the raid had stomped them out. It was, she told the *Earth First! Journal*, "the most disgusting abuse of power I have ever seen." Just four months before, she had enough faith in the system to go door to door for the Sierra Club. Now she had seen the state responding to unarmed protesters with a paramilitary operation. "The authorities came in with the intent to hurt people. I believe that if they had the chance they would have had no problem killing people," she said. Natalia, Dr. Toxic, and others who were in the houses that night had nightmares for months after the raid. The sound of a helicopter gave Natalia flashbacks, and she would start to shake and cry.

Within days of the raid, activists from the camp and their supporters issued a statement demanding "a full independent investigation into charges of brutality, torture, and use of excessive force by government police." But with no outside witnesses, it was their word against the authorities.' The media, much to the activists' disgust, had not been inside the basements.

Only one news team had gotten word of the raid ahead of time. That team went to the airport and was allowed to ride along in a police bus behind the Ryder trucks, Inspector Indrehus recalled. What the police didn't tell the news team was the bus's destination—the back perimeter of the site, far from the houses. Reporters who got to the scene soon after police sealed off the camp fared little better. They had two choices: to cover the story as best they could from behind the police tape, or get arrested for crossing it, spend the night in jail, and miss the rest of the story. Only one member of the press managed to get behind the police tape. Dick Bancroft, seventy-one, a freelance photographer and father of the polar explorer Ann Bancroft, was inside the camp when the Ryder trucks arrived. He took a few flash photos before police knocked him to the ground, arrested him, and confiscated his camera. It was returned to him three weeks later, minus the film.

When the raid was over, one woman claimed that her nose had been broken by a pain compliance hold, but in her arrest photo her face did not appear bruised or swollen. A man said police had banged his head on the step as he was dragged out of a house, but his arrest photo showed no cuts or bruises. Neither did the arrest photos of the faces of any of the thirty-four adults apprehended in the raid.

At a press conference two days after the raid, the activists said they had been pepper-sprayed and denied medical attention, but their assertions of mistreatment sank with hardly a ripple. "The allegations they are making are without merit," a spokesperson for the Department of Public Safety told the Minneapolis *Star Tribune*. "All law enforcement officers acted in an exemplary manner."

It didn't help the activists' credibility when they used words like *tortured* and *brutalized* to describe what had happened. One middle-aged activist was especially outspoken in his insistence that he'd been "tortured" when an officer put pepper spray in his face, but the rest of his story was an anticlimax. "It was like my whole face was severely sunburned," he said in a faltering voice. "My lips were so chapped ..." When all the evidence was added up, it seemed that Operation Coldsnap, however frightening in its scope, had been more inept than brutal.

But the activists did win a measure of vindication months later, when Hennepin County district judge Stephen Aldrich dismissed the charges against Natalia and Dr. Toxic on the grounds of police misconduct. Aldrich ruled that the officers' use of pepper spray had been acceptable, but that other conduct had not been. The officers had refused to provide medical attention for Dr. Toxic and Natalia until they gave their real names, which they refused to do. (The ever-defiant Natalia identified herself as No Compromise. Asked for her full name, she had answered, "No FUCKING Compromise.")

"On the record here, the lack of medical attention and symptom relief provided to these defendants is unacceptable," the judge ruled. He noted that Natalia and Dr. Toxic had suffered "extreme pain, blindness, vomiting, uncontrollable crying, and drooling.... Not only were the defendants in physical agony for the hours they were in custody, but the aftereffects of the irritant lasted for days. What makes the situation even more outrageous is that medical attention was accessible, but not dispensed." The ruling concluded, "The conduct exhibited here was outrageous, especially considering the nature of the crime charged. For the court to turn a blind eye at the point of prosecution to the defendants' pain would render our system no better than that of countries we criticize for state-sanctioned torture.... It would be unconscionable to add a criminal trial to the pain these defendants suffered on the street."

But by then it was August and the story just an epilogue. On an eighty-five-degree day in Minnesota's brief northern summer, few people gave much thought to something that had happened so long ago in the cold and snow and dark.

Much later, MnDOT's Bob McFarlin and safety commissioner Don Davis would remember Operation Coldsnap with the relief people feel when a catastrophe has been averted. "Nobody got hurt," said McFarlin. "Nobody got hurt, and that was the goal."

"You ask yourself, how many were killed? How many were injured? How much property was damaged?" Davis said. "Zero, zero, zero."

Minneapolis police inspector David Indrehus was also relieved that nothing major had gone wrong. He had been in a tough spot, holding a job just high enough that he would have to take the blame for mistakes, but not high enough to give him real control over decisions. He remembered standing on the airport tarmac the night of the raid, in front of hundreds of officers who had arrived from all over the state. He had explained who the protesters were and why they were there, but he knew in the end that it didn't matter. "Working cops, they just have to know what their job is. That was why we were all there. The police department, the State Patrol, the sheriffs, they always kind of get the blame for a lot of stuff, but we basically became pawns," he said. "We were told, in this case by the politicians, do this and make it happen. I could have had 50 percent of those cops there that agreed with the protesters, and it wouldn't have made a difference."

He remembered telling the officers, "Here's what we're going to do. You folks are getting on the buses, you folks are taking your cars. Load up, and let's go."

Indrehus retired at the age of fifty-one, two weeks after the raid. He was fit and lean, with a clipped gray mustache and a military bearing, but the stress was getting to him. He oversaw 170 officers in a precinct with the highest crime rate in the city and two or three homicide calls a week. One Friday, Indrehus's partner of thirty years learned he had cancer. By Wednesday he was dead. "I said, 'That's enough. Life's too short,'" Indrehus recalled later. He would see Operation Coldsnap to the end, then get out. "That was my swan song," he said.

In the gray morning after the raid, bulldozers and backhoes made a steady low roar punctuated by the cracking of boards and the dull boom of walls tumbling into basements.

Tumbleweed sat in the falling snow, tears running down his checks and freezing in his fuzzy beard. "The snow was just piling up on me. At that moment I felt like it was over. This was the end of everything we'd worked for," he said later. Snow soon hid the foundations, covering every sign that houses had ever been there.

The old oaks that shaded the neighborhood still stood. Within hours, what had once been a city street looked like a wooded riverbank. Someone had stapled pink paper hearts to the big oaks' furrowed trunks. The snowflakes drifted down on a little doomed forest, through the stout crooked branches standing dark against the pale sky.

About an hour before midnight on Christmas Eve, three cars left the Walgreen's parking lot on Hiawatha Avenue, drove seven blocks south, and parked near the field by Riverview Road. Bill Busse, his camp friend Garrett, and a handful of others got out cautiously, watching for cops, then sneaked through the snow toward the four bur oaks. Garrett chained himself by the waist to the easternmost tree. Only then did they build a fire.

The temperature was zero, so cold that the snow was a dry powder that squeaked when they walked in it and it was impossible to stay warm standing still. They stiffened and hunched their shoulders and swung their arms and shifted from foot to foot. They piled blankets over Garrett and dragged a mattress out from the car for him to sit on, but the chain was so cold it seemed to suck away any warmth. Garrett couldn't stay locked down for long before someone had to take his place, then someone else after that, all night long.

But Bill Busse felt better than he had just a few hours earlier, in the warm coffee shop where he and Garrett had hatched their plan. "We weren't giving up. We weren't quitting," he said later. They had called Jim Anderson and asked what he thought of a new camp and he said fine, go ahead.

For days, Bill had been brooding over the raid. He realized only afterward that there was no documentation of what happened in the basements that night. He and his comrades had assumed that photographers would manage to get inside the police lines. The day after the raid, he had stood gloomily in the snow where the camp had been, letting the realization sink in: no photos. It was an incredible mistake, he now knew, a total tactical error. He had searched through the debris and found a tent pole and a soggy sleeping bag. He fumed at the thought of the police warming their hands over people's burning belongings. Now, on Christmas Eve, four days after the raid, he was fighting back.

"It was something an individual could do, and it made you powerful," Bill explained later. "The state wasn't just going to do

Garrett chained himself to the easternmost of the four oaks in the route of Highway 55. He told the newspaper his name was George Hayduke — one of the ecosaboteurs in Edward Abbey's novel *The Monkey Wrench Gang*. Staff photograph by *St. Paul Pioneer Press*; copyright *St. Paul Pioneer Press*.

this, bring in their eight hundred officers and just be done with this. It wasn't going to be that easy for them."

When the long night was over and Christmas Day dawned with no cops in sight, the activists set up the first tent. Then a village made of tents, tarps, and scrap lumber rose near the four trees. Another canvas tent, reminiscent of the one at Camp Two Pines, went up, sheltered by a blue plastic tarp that soon sagged under new snow. Dome tents like soft igloos clustered nearby. Jim Anderson borrowed a decrepit motor home and parked it next to an ancient, bright-blue school bus. People scrounged metal cabinets and a bottled-gas stove for an outdoor kitchen, and they ringed the campfire with a windbreak made of straw bales and plywood.

The new camp was smaller and quieter than the old one. Some of the Earth First!ers lived there on and off, and Jim Anderson spent a couple of nights a week, but it was so cold he didn't expect a lot of regulars. A few watchdogs were enough for now. As the days went by, Jim and the others began to think the Department of Transportation didn't plan to cut the trees anytime soon. After all, months of bitter cold stretched between then and the construction season next spring. But you never knew.

On a bleak day in early January, two activists called Bear and Caleb stood silently by the fire, small figures outlined against the field's expanse. Wherever they moved on the tromped, muddy ground, wind blew smoke in their faces. It carried the caws of crows and the distant whoosh of tires in the slush of Hiawatha Avenue. It fluttered the scraps of red, yellow, black, and white cloth tied in the branches of the four oaks behind them, and it stirred the strings of beads and the twist of braided sweet grass looped around the gray, furrowed trunks. At the center of the diamond formed by the four oaks, traces of offerings lingered: a candle, a pile of stones, a feather.

Asked about the raid, Bear explained that he was inside the lodge pounding a drum when the police came. He said the sacred drum was smashed, and people's clothing and sleeping bags were loaded into a dump truck and taken to a landfill. There was a long pause while everyone stared at the fire.

"It's quieted down with me since the raid. How about you, Bear?" said Caleb suddenly. An intense young man with a sparse beard and a kerchief pulled over his forehead, nearly hiding his eyes, Caleb spoke with an odd affect that always made strangers look twice at him. Bear, silent, kept his eyes on the fire.

Nearly sixty, strong and sturdily built, Bear was a Northern Cheyenne who had grown up scarred by his years at an Indian boarding school but seldom spoke of it. He had worked as a security guard for the American Indian Movement for years, patrolling on Minneapolis's toughest streets. He wore a bear claw necklace and an army jacket with a constantly changing collection of buttons. "American Indian Movement," they said, or "Free Leonard Peltier"—the long-imprisoned AIM activist.

Bear was taciturn with outsiders, but he had befriended many of the young Earth First!ers, who looked to him with respect. He didn't spend a lot of time talking, but when he did, people listened. Today, though, he wasn't saying much.

A long pause, then Caleb said, "I've told friends and family what happened—anyone who makes the mistake of calling people protesters." He said the campers should be called "peaceable assemblers." He walked over to a cardboard sign that read "We are committed to non-violence."

"See that?" he said. "That's new, new to these times."

The silence stretched out. Asked how things were going now, Bear said they needed wood, and that it took about a cord a week to keep the sacred fire going.

Were the days long?

"Sometimes it is, some days it goes OK," Bear said. Sometimes they went to coffeehouses to warm up, he added after a pause. "You just have to take it from day to day. Hopefully we're going to win. We just have to see now."

Caleb began explaining the red, yellow, black, and white cloths on the four oaks. The colors symbolized the four colors of man, he began, then broke off suddenly. "You ever have to deal with someone who dies? You know what it's like? I haven't." He turned away abruptly.

Behind him in the field, Bear stood stolidly by the fire.

On another day, a round-faced girl called Tarzana, eighteen, warmed her hands by the flames. Besides Bear and Caleb, she was the only other person living full time in camp. She had graduated from a small-town high school in rural Minnesota, moved to the Twin Cities, and heard about the Free State. She was arrested in the raid and still had nightmares in which all was dark and peaceful, then everything happened at once. "You fear for your life, and you fear for the life of your friend," she said.

She wore a giant army-green parka and a black wool stocking cap pulled low against the cold. Her driver's license showed a younger, thinner girl looking up from under her bangs like Princess Diana. Her real name was Anne-Marie Gabrielle Rush. Her mother had been unhappy when she heard from news reports that the protesters had tapped into the gas lines; she insisted that the governor wouldn't lie about such a thing. At the thought of her mother's confidence in the governor, Tarzana gave a short sarcastic laugh.

She had dreams for the new camp. "We're in the process of remodeling," she said playfully. "You'll see tipis, tents.... If we're still here in the summer there will be so many people it will be unbelievable." Someday, she said, this field would all be gardens. But summer was many frigid months away, and for now, the action had shifted away from the camp.

Jim Anderson's relatives, led by his uncle Bob Brown, wanted to take a different approach. They were tired of Earth First! and its tactics. They had seen where working outside the system had gotten the protesters, and they wanted to try a different tack.

A soft-spoken, fifty-eight-year-old housepainter with hazel eyes and thinning curly hair pulled back in a ponytail, Bob Brown led a group of families who called themselves the Mendota Mdewakanton Dakota Community. Lacking federal recognition as a tribe, they began to organize in 1996, when nearly three hundred members of Bob Brown's extended family decided to apply for federal recognition and chose Bob as chairman and Jim as cultural chairman. From then on, Bob Brown nearly always wore a red warm-up jacket with "Mendota Mdewakanton Dakota Community" embroidered on the back.

He bitterly regretted the day his family had gotten involved with Earth First! It had made them look like radicals, he thought, at a time when they were struggling for federal recognition and for acceptance by other recognized tribes. The Mendota Dakota were ordinary, mainstream citizens, except perhaps for his two hot-headed nephews, Jim and Michael. His people had chosen the wrong tactics, Bob Brown realized unhappily. Now the public wondered who the Mendota Dakota were and why, if the land the highway would pass through was really so sacred to them, they hadn't said anything about it decades earlier. It wasn't easy to explain.

The answers lay deep in the family's history. It was an ambiguous and not always proud heritage, a long and complicated tale with pieces missing and parts hidden for years by generations who didn't talk about the past. It began in Mendota.

CHAPTER 4

Little Crow's Children

The town of Mendota, population 167, is a wide spot on the highway that runs atop the Mississippi River bluffs just a mile or so south of Minneapolis. Tiny old wooden houses and modest newer ones cling to wooded hills that form a semicircle above the town. The main street is lined with a metalworking shop, a post office, a supper club, and a VFW lodge with a sign saying "FISH FRY FRIDAY NIGHTS." It takes a minute or two to drive the town's entire length, before the road sweeps up a hill and into the tangle of freeways leading to other, busier places. But Mendota's main street does have something worth stopping for. The signs on the freeway label them "Mendota Historic Sites." The graceful, weathered brick and stone buildings are the remains of an old fur-trading post, a gathering place for the branch of Dakota known as the Mdewakanton, the easternmost tribe of what was once called the Great Sioux Nation.

In the Dakota language, *Mnidote* means the mouth of a river. Mendota overlooks the confluence of two rivers, the Mississippi and what is now called the Minnesota River, once named the St. Pierre by the French Canadian fur traders who reached it by canoe after 1700.

"The River St. Pierre ... falls into the Mississippi from the west," wrote Jonathan Carver, a fifty-seven-year-old New Englander who explored the area for Great Britain in 1766. "This is a fair large river that flows through the country of the Sioux, a most delightful country. Wild rice grows here in abundance. At a little distance from the river are hills from which you have beautiful views."

In September 1805, a seventy-foot keelboat poled by U.S. Army privates reached the spot. The expedition, led by First Lieutenant Zebulon M. Pike, made camp at an island that now bears his name. The young officer had embarked from St. Louis that summer, under orders "to proceed up the Mississippi with all possible diligence." Pike was to follow the river north to its source, noting its topographical features and the number of resident Indians. "In addition to the preceding orders," Pike's instructions continued, "you will be pleased to obtain permission from the Indians who claim the ground, for the erection of military posts and trading houses, at the mouth of the river St. Pierre, the Falls of the St. Anthony [on the site of present-day Minneapolis], and every other critical point which may fall under your observation."

On the island, Pike met with a party of 150 Dakota. Among them was the Mdewakanton leader Cetanwakanmani, or Sacred Hawk Walking. The French voyageurs called him Le Petit Corbeau—Little Crow—because of the crow wings dangling from his belt. Pike drew up a treaty granting the United States sovereignty over the Mississippi and its banks for a nine-mile stretch extending from below the confluence of the two rivers north to St. Anthony Falls. With two strokes of a quill pen, Little Crow and another leader gave their consent. "We, the undersigned, have hereunto set our hands and seals, at the mouth of the River St. Peter's.... Le Petite Corbeau, his X mark, Way Aga Enagee, his X mark." In return, Pike distributed two hundred dollars worth of trade goods and sixty gallons of whiskey.

By 1820, U.S. Army soldiers were erecting what the Mdewakanton must have thought an unimaginably large building. From a cliff where the two rivers met, Fort Snelling dominated all the surrounding countryside. "[The fort] is constructed of stone ... and being placed on a commanding bluff, has somewhat the appearance of an old German castle, or one of the strongholds on the Rhine," wrote Mary Eastman, the wife of an army officer stationed there. She noted that it was "one of the strongest Indian forts in the United States."

Whatever Cetanwakanmani may have wondered as he saw the stone fort rising, he had had little reason to be wary of whites before then. His village had prospered after the white men arrived. Fur traders like Joseph Renville and Jean Baptiste Faribault had married into his family and established a trading economy knit together by kinship. It was the beginning of a new, mixed-blood society intimately tied to the traditional Dakota.

"The French Canadians, who are here employed by the Fur Company, are a strange set of people," wrote a British naval officer who visited Fort Snelling in 1838. "Occasionally they return to Canada with their earnings, but the major part have connected themselves with Indian women, and have numerous families; for children in this fine climate are so numerous, that they often appear to spring from the earth."

When the Catholic frontier bishop Mathias Loras traveled to Mendota the next year, he found a thriving mixed-blood settlement. "To his great astonishment he found there not far from the fort 185 families, consisting mostly of Indians or French," a Dubuque newspaper reported. "No pen can describe the joy which this apparently lost flock of the Church manifested, when its members saw this bishop in their midst." Indian women were "favourably disposed towards religion," Bishop Loras noted. "We baptize a great number of children." Three years after the bishop's visit, the people of Mendota dedicated a small stone church on a hill overlooking the river valley. St. Peter's Church archives trace the births, deaths, and baptisms of generations of mixed-blood families.

Baptismal records show a little girl named Lillian Felix, born in Mendota in 1881. St. Peter's Church and other records track her family tree through French and Indian branches stretching back more than a century. They show her grandmother, Rosalie Freniere, born Mazasnawin sixty years earlier; further back are Lillian's great-great-grandparents O-Mon-Dah-Gah-Ne-Nee (John Baptiste) Bellecourt and his wife, Sah-gah-Ie-way-quay (Everlasting Woman); at the very furthest reaches are Sah-gah-Ie-way-quay's parents, No-din-ah-quah-um (Wind Rising), born in 1760, and We-go-baince, born in 1770.

Church records also show that Lillian's husband, Albert LeClaire, had among his Indian and French Canadian ancestors the mixed-blood trader Joseph Renville and his Dakota wife Mary, granddaughter of Cetanwakanmani, Sacred Hawk Walking.

Albert LeClaire was born in Mendota in 1885. Like his family tree, his birth certificate provides evidence of his mixed-blood heritage, albeit less poetically. In the blank for "race," it says "red and white."

Albert LeClaire and Lillian Felix were married in 1905. They were Bob Brown's grandparents.

Growing up in South Minneapolis, Bob Brown knew he was part Indian, but his mother, Selisha, downplayed the family's Native American heritage. Bob's older sisters remembered her telling them they had so little Indian blood, it would all be gone if they

The grandparents of Mendota Mdewakanton Dakota Community chairman Bob Brown: Albert LeClaire and Lillian Felix LeClaire, shown with their son Albert. Photograph courtesy of Mendota Mdewakanton Dakota Community.

pricked their fingers. If they were asked, older family members would say they were mostly French.

"It was a deep dark secret they had Indian blood," Bob's wife, Linda Brown, remembered. "They didn't want their kids to know. Bob's uncle, even now he doesn't want to believe he's Indian."

But Bob Brown could tell by their appearance that members of his family had more than a drop of Indian blood. His uncles all had dark skin, black hair, and Indian features. And he saw others like them when the family went to visit his grandmother Lillian at Mendota. "I knew that Indian people had lived there forever," he said.

Bob Brown was in his fifties before he tried to trace the history of the mixed-blood community he came from. It wasn't easy. The Felixes, LeClaires, Renvilles, and other French and Indian families who had lived in Mendota when it was a thriving fur trade post had seen hard times since then. They had been swept and scattered by winds that transformed all of Dakota society, full blood and mixed blood alike.

The spring of 1862 followed a hard winter, including a February that was one of the coldest in all of the nineteenth century. Deep snows fell along the Minnesota River valley, where Minnesota's entire Dakota Nation now lived. Ever more restrictive treaties had confined them to a fraction of their former lands: a ten-mile-wide strip along a stretch of the Minnesota River, which flows for about a hundred miles through the southwestern part of the state, from the South Dakota border to just above New Ulm. The snow had come early and lingered into April. By spring, many of the Dakota were hungry. Banned from their old hunting grounds, they depended on food and payments the government had promised in exchange for their land, but those payments were often slow in coming, stolen by corrupt agents, or seized by un-scrupulous traders.

In August, Cetanwakanmani's grandson, also called Little Crow, and several hundred Dakota met with a government Indian agent and four traders who owned warehouses on the reserva-tion. "We have waited a long time," Little Crow said. "The money is ours, but we cannot get it. We have no food, but here are stores, filled with food. We ask that you, the agent, make some arrangements by which we can get food from these stores, or else we may take our own way to keep ourselves from starving. When men are hungry, they help themselves." Trader Andrew Myrick gave this answer: "So far as I am concerned, if they are hungry, let them eat grass or their own dung."

Two days later, four young Mdewakanton Dakota men look-ing for food killed five white settlers: three men and two women. Shakopee, the leader of the band that included the four young warriors, went to Little Crow and told him what had happened. "He said war was now declared. Blood had been shed, the payment would be stopped, and the whites would take a terrible vengeance because women had been killed," another band leader, Big Eagle, remembered. "Wabasha, Wacouta, myself and others still talked for peace, but nobody would listen to us, and soon the cry was 'Kill the whites and kill all these cut-hairs who will not join us.' A council was held and war was declared."

Little Crow warned the young men about going to war against the whites; he had ridden steamboats and trains to Wash-ington, D.C., and seen for himself the unimaginable hordes of white people who lived east of the Mississippi. "See!—the white men are like the locusts when they fly so thick that the whole sky is a snow-storm. You may kill one—two—ten; yes, as many as the leaves in the forest yonder, and their brothers will not miss them," he said in a speech remembered by his young son. But

Little Crow, whose Indian name was Taoyeteduta, reluctantly agreed to lead the war. "You will die like the rabbits when the hungry wolves hunt them in the Hard Moon (January)," he said. "Taoyeteduta is not a coward. He will die with you."

As Big Eagle recalled, "Parties formed and dashed away in the darkness to kill settlers. The women began to run bullets and the men to clean their guns." Early the next morning, warriors attacked the reservation's southern headquarters and trading post. Among the dead was trader Andrew Myrick. "He said to them: 'Go and eat grass,'" Big Eagle remembered. "Now he was lying on the ground dead, with his mouth stuffed full of grass, and the Indians were saying tauntingly: 'Myrick is eating grass himself.'"

By fall, some six hundred white settlers and uncounted Dakota were dead. Farms across southeastern Minnesota lay abandoned. As the war ended, troops led by Henry H. Sibley, the former head of the fur-trading post at Mendota, drove the Dakota warriors from the state and pursued them and their families westward across the plains. Some fled to Canada, but many were rounded up and sent to a barren stretch of land along the Missouri River in what is now South Dakota.

In December, after cursory trials, thirty-eight Dakotas and mixed bloods were hung from a single huge scaffold at Mankato, Minnesota, in the largest mass execution in the nation's history. Only the intercession of President Abraham Lincoln prevented the execution of 266 more.

Little Crow was shot and killed by a white farmer. The corpse was later scalped to collect the seventy-five-dollar bounty offered by the State of Minnesota for hostile Dakota Indians. In addition, the state legislature awarded the farmer a special bounty of five hundred dollars. Little Crow's skull, scalp, and forearms were put on display in a case at the Minnesota Historical Society's museum in St. Paul. They remained there till the early 1900s, when museum officials quietly removed them to a storage shelf.

Little Crow had led the Dakota in a desperate attempt to save their traditional way of life, and they had paid a terrible price. Virtually the entire Minnesota Dakota Nation was now in exile. Only those Dakota who had sided with the whites were allowed to remain in Minnesota. They became the ancestors of the present-day Mendota Dakota.

Those who stayed included mixed bloods from the old Mendota fur post and the so-called farmer Indians, families who under the influence of Christian missionaries had given up hunting for farming and abandoned their traditional religion for Christianity. As the outward sign of their conversion, the farmer Indians adopted settlers' clothing and the men cut their hair.

"It was forced on them. I don't blame anybody to this day," Jim Anderson said when he was asked about his ancestors' assimilation and role in the war. "Their way of life was taken from them and they had to change. So it's not something that I look back on that they did on purpose. I'm not proud of some of the things that some of 'em probably did, but I still don't blame them."

But like their traditional relatives, the assimilated Dakota paid a bitter price for the war. Many had been taken prisoner by the warring Dakota and held in camps along with white captives. As the war ended, the "friendly Indians" helped release the white captives unharmed to Henry Sibley's troops. In return, they hoped to be able to take up their lives where they had left off. They were soon disappointed.

In November 1862, a wagon train that at times stretched four miles long straggled along the Minnesota River valley. Soldiers rode on both sides, escorting a ragged column of Dakota and mixed-blood people on a forced journey to Fort Snelling.

George Crooks, the six-year-old son of Christian farmer

Indians, rode crammed into an oxcart with his sixteen-year-old brother and two Indian men. "We were bound securely and on our journey resembled a load of animals on their way to market," Crooks told a newspaper reporter years later. "We traveled slow, meeting now and then a white person who never failed to give us a look of revenge as we jolted along in our cramped condition."

As they approached the southern Minnesota town of New Ulm, George's brother told him the driver was afraid to go through town. "Overcome with fear," the little boy crouched down beside his big brother. "Women were running about, men waving their arms and shouting at the top of their voices.... [It] convinced the driver the citizens of that village were wild for the thirst of blood, so he turned the vehicle in an effort to escape the angry mob, but not until too late, they were upon us. We were pounded to a jelly, my arms, feet and head resembled raw beef steak. How I escaped alive has always been a mystery to me.... My brother was killed and when I realized he was dead I felt the only person in the world to look after me was gone and I wished at the time they had killed me."

Another who made the journey, Good Star Woman, remembered the terror: "The soldiers rode on each side of the column of Indians and tried to protect them but could not always do so." Her father hid her and her two little sisters under a buffalo robe, but peeking out from under it she saw townspeople attacking women and children with poles, pitchforks, and axes. "At length the pitiful column of friendly Sioux reached Fort Snelling," she remembered. "A high fence was put around their camp, but the settlers came and took their horses and oxen."

"The soldiers drove a wagon among the tents and gave crackers to the children and bread to the older people," Good Star Woman remembered. "Measles broke out, and the Indians thought the disease was caused by the strange food.... Sometimes

20 to 50 died in a day and were buried in a long trench, the old large people underneath and the children on top."

There is no official record of the death toll, but in December 1862, a federal "Census of the Indian Camp" showed 1,601 inhabitants. The month before, the *St. Paul Pioneer Press* had reported the arrival of more than eighteen hundred. If the two figures are correct, and assuming none of the captives was allowed to leave between November and December, the death toll in those few weeks would stand around two hundred. William Folwell's four-volume *History of Minnesota*, a standard reference work, devotes one line to the subject: "There was some sickness and there were not a few deaths."

A photograph of the Indian camp taken that winter shows tipis on the river flats below Fort Snelling, in military rows surrounded by a high wooden fence. A cold mist has settled on the river bottom, swirling white and ghostly around the dark tipis.

Dakota tipis in an internment camp on the river flats below Fort Snelling during the winter of 1862–63. Photograph by Benjamin Franklin Upton; courtesy of the Minnesota Historical Society.

When the winter ended, most of the camp's remaining in-habitants, primarily women and children and a few old men, were crammed into steamboats and sent downriver to join the other Dakota in exile on the reservation in South Dakota. At the St. Paul landing, a mob threw stones onto the crowded decks.

When the steamboats departed, fewer than two hundred people were left at the Indian camp. They included the families of mixed-blood scouts who were out helping soldiers track the defeated Dakota across the plains. The scouts' leader was Gabriel Renville, a descendent of the trader Joseph Renville who had married into Little Crow's family long ago at Mendota.

A small band of farmer Indians also lingered at the camp. They had cast their lot with the whites; some had saved white captives, and some had even testified in court against the Dakota who were hung at Mankato. Now they were afraid to join the rest of the Dakota on the reservation. That winter, Taopi, or Wounded Man, the leader of the farmer band, and other farmer Indians had sent a petition to President Lincoln, pleading for land of their own. "We are farmers, and want that our Great Father would allow us to farm again whenever he pleases, only we never want to go away with the wild blanket Indians again; for what we have done for the whites they would kill us," they wrote. "We think we have not forfeited our annuities or other funds, because we have done no wrong.... We must have food and clothing, and in the spring somewhere to live."

Minnesota's first Episcopal bishop, Henry B. Whipple, pleaded the farmer Indians' cause. Although he was unable to obtain land for them, he arranged for a temporary refuge in southeastern Minnesota near the town of Faribault. Alexander Faribault, of the old fur-trading family, agreed to let them stay on land he owned just outside the town.

Faribault soon had to defend himself against a flurry of

rumors. "Having been informed that a report is current that I am harboring guilty Indians, and that there are now at my place a large number, some of whom are known to have participated in the outbreak, and that threats of violence to any Indians found there have been made, I deem it my duty to quiet the fears of persons who might believe such report to be true, though I hope my fellow citizens will examine for themselves," he wrote to the local newspaper in June 1863. "I know these Indians well, and I know them to be harmless, innocent and good persons." He carefully listed each family. The camps inhabitants included Taopi and his mother, Berry Picker; a woman identified as the wife and mother of Good Thunder, who was "now employed as a scout for Gen. Sibley"; and "Wacon, or La Clare and his family, *who were here during the outbreak,* and are known to be entirely innocent." Wacon La Clare was Bob Brown's great-great-grandfather.

At Faribault, the families lived in poverty. They raised a few crops and dug ginseng from the woods, but mostly depended on Alexander Faribault's charity. Bishop Whipple tried in vain to get the government to grant them farms of their own. "I have plead[ed] for this poor race until I am heart sick," he lamented, but still Taopi and the other families remained in limbo. "If they are still alive, it is by the mercy of God," Whipple wrote bitterly.

After several years of near starvation, some of the families at Faribault agreed to leave Minnesota for the reservation. The rest still refused. They moved instead to a much more familiar place. Henry H. Sibley, the onetime fur trader who had led the troops against the Dakota, agreed to let the farmer Indians stay on some of his land. It was on a hill overlooking the confluence of the Minnesota and Mississippi Rivers, on the outskirts of Mendota.

Old maps of Mendota show plots labeled "H. H. Sibley's Indian Homes" but beyond that the historical record is scant. A St. Paul newspaper, the *Globe*, mentions them in passing in an

1887 travel article about the quaint backwater of Mendota. It states that Sibley intended the plots as a place "to gather together the scattered families of his old Indian scouts, where the children could have the advantages of school and church and be taught the ways and customs of whites." The article says that "only four or five families availed themselves of the opportunity," but gives no indication if these were scouts, farmer Indians, or both.

St. Paul newspapers reported the comings and goings of Taopi's aged mother, Berry Picker, who made trips from Mendota to beg on the streets of St. Paul. Berry Picker, who had grown up just a few miles from the city in Little Crow's village, was well known in St. Paul, "where she always had a kind smile for

Berry Picker, one of the Dakota who took refuge at Mendota after the Dakota Uprising of 1862, begging on the streets of St. Paul, ca. 1868. Photograph by Whitney & Zimmerman Photographers; courtesy of the Minnesota Historical Society.

everybody," wrote a reporter for the *St. Paul Dispatch*. "Since the Sioux war the historic old squaw has often appeared in our streets, with her long cane, bowing and smiling and begging, and usually she has carried back to Mendota, where she resided on property owned by General Sibley, the kind offerings of her friends, the whites."

A photograph of Berry Picker in the Minnesota Historical Society archives from 1868 shows an old woman dressed in a ragged cloak, resting on a packing crate in a snow-covered alley. Her face is etched with deep lines. She leans on a stick, a burlap sack at her feet.

Other historical records also give hints of the Indian community at Mendota. Photographs from the 1880s show families living in tipis in the woods outside town. In one, labeled "Mendota Sioux Girl, ca. 1880's," a child in a calico dress looks solemnly at the camera. She stands with her chin on her chest, the

Dakota girl outside a tipi at Mendota in the 1880s. Photograph courtesy of the Minnesota Historical Society.

sole of one bare foot pressed awkwardly against her ankle, her arms clutching a rag doll. Behind her, the door to a worn canvas tipi is folded open, revealing the everyday objects of her family's life: mats and pillows, a pair of moccasins, a wooden box, a china dish.

The *Globe*'s 1887 travel article includes a sketch of tipis along Lake Augusta just south of Sibley's Indian homes. Reporters who visited Mendota's quiet main street noted its mixed-blood community: "There we find a few straggling teams, a pretty brunette half-breed, some worthy Frenchmen," the *Globe* reporter wrote. Another contemporary article, headlined "Mendota the Beautiful," plugged the town's quaint attractions, including its inhabitants. "A dark eyed little lad, whose color betrays the Indian blood, will bring you water in a pail," it reads, "and if you toss him a dime, he will be too astonished even to thank you."

In 1902, a *Globe* reporter visited Mendota and described "a queer conglomeration of frame houses, huts, and earthen hovels" that made up the tiny Indian community. "At Mendota there still lingers the forgotten remnant of a dying people where in a portion of the town, apart from the rest of the inhabitants, two or three families forming a settlement of their own is all that now remains of the once great Sioux nation. . . .

"Off by themselves in a little hollow backed by the hills that form an almost perfect semi-circle and known locally as the 'Horseshoe,' live a handful of Indian and halfbreed families in half a dozen houses forming the settlement. . . . The pure Indian has small place here, but the halfbreed, the quarterbreed and every other fraction of French and Indian blood is to be found. . . .

"Time was when all this country round was inhabited by Indians," the reporter noted, but "this melancholy little group is all that remains." The village included "four families, descendants of an Indian called Peter Felix." Those descendants included a young woman named Lillian Felix, Bob Brown's grandmother.

A reproduction of Lillian Felix's portrait, painted around that same time, hangs on Bob and Linda Brown's living room wall. Lillian's descendants believe the portrait was done when Lillian was back east at the Carlisle Indian Industrial School, a Pennsylvania boarding school run out of a former military barracks and dedicated to assimilating Indian children. In the painting, Lillian is wearing a dress with a lace collar pinned with a brooch. Her brown hair is pulled back with a velvet ribbon. Her dark eyes gaze steadily at the painter. Her lips are pressed firmly together, not quite smiling.

Bob Brown's older sister, Lillian Brown Anderson, nicknamed Mickey, remembers her grandmother Lillian well, even though Mickey was only six years old when Lillian died. Mickey practically lived at her grandmother's white frame house in Mendota, just off the horseshoe-shaped hollow in the hills. "She was soft, smelled good, baked all the time," Mickey recalled.

Portrait of Lillian Felix, believed to have been painted while she was a student at the Carlisle Indian Industrial School, a Pennsylvania boarding school dedicated to assimilating Indian children. Photograph courtesy of Mendota Mdewakanton Dakota Community.

She remembered her grandmother heating water on a wood stove in a big copper boiler. Horse teams hauled the water from the bottom of the hill, using wagons in the summer and sleds in the winter.

Mickey remembered Mendota as a small town where everybody knew everybody, where people heard mothers calling for their children at dinnertime. People went every day to the post office for their mail. "Everybody would be there like hens talking," Mickey said. There were LeClaires, Felixes, LaCroixes, and the extended Robinette family, so big it made up half of Mendota. Everything revolved around St. Peter's Church, where Mickey went every Sunday, learned her catechism, and was confirmed. When she was ten, her family moved to St. Paul and came back to Mendota only for visits.

Years later, when one of her distant cousins started digging into the family's roots for a college genealogy project, Mickey didn't pay much attention. Then one Easter Sunday, Mickey's son Jim Anderson brought the papers with him to dinner at Bob and Linda Brown's house. "Jim said, 'Mom, you've got more Indian in you than you think you do,'" Mickey remembered. "He said, 'Look here, Grandma was a lot more Indian than we knew.'" Jim showed her the papers; they said Lillian Felix was half Dakota and one-quarter Chippewa. "I said, 'She is? Well, for heaven's sake.'"

Mickey let the kids do the digging. They had always heard that Lillian had gone to school back east. Now they learned it had been an Indian boarding school. Mickey's mother, Selisha, had never mentioned it. Mickey realized now how closed-mouthed her mother had been. Mickey thought back to when she was little and her uncles came to visit. "I used to say to my mother, 'Boy are they ever dark,' and she'd say, 'Oh, they work outside,'" Mickey said.

When Albert and Lillian LeClaire's son Albert Jr. was dying, he told stories about growing up Indian—stories that he kept hidden for most of his life. ("I would give anything on earth to be able to do that with my mom," Mickey said. "I'd say, 'Mom, there's nothing to be ashamed of. Just tell us.'")

Albert and Lillian's son revealed unhappy memories of the years the family had lived away from Mendota, on an Indian reservation. Albert and Lillian owned a seventeen-acre farm on what was then known as the Prior Lake Indian Settlement, near the town of Shakopee, ten miles southwest of Mendota in the Minnesota River valley. The family moved to the farm in 1919. They had taken the federal government up on its offer of farmland for Mdewakanton Dakota who could prove they lived in Minnesota in 1886—twenty-four years after the end of the Dakota war. Only a few hundred people met those qualifications.

Albert LeClaire on his farm at the former Prior Lake Indian Settlement. The settlement later became the Shakopee Mdewakanton Dakota Reservation and the site of the Little Six bingo parlor. Photograph courtesy of Mendota Mdewakanton Dakota Community.

Life on the reservation was much harder than in Mendota. In the town of Shakopee, where the five LeClaire children went to school, they were called "half-breeds" and "dirty Indians." The two youngest children would come home with cuts and bruises. Lillian complained to the teacher and was told that her children just needed to toughen up. Finally, she and the younger children moved back to the old white frame house in Mendota.

Her husband, Albert, stayed behind at Shakopee, still hoping to succeed as a farmer. In the fall of 1937, at the age of fifty-two, he applied to the commissioner of Indian affairs in Washington, D.C., for more land on the Prior Lake Indian Settlement. He hoped that with a total of forty acres he would have enough to make a good living. On the form, he replied yes to a question asking if he was a Mdewakanton Sioux Indian, and he gave his father's and mother's "degree of Indian blood" as one-quarter and one-half, respectively. The next spring, he learned that his application had been accepted. The land was his.

But Albert never did make it as a farmer. Four years later, he was thrown from a car that overturned in the ditch near his farm. The hospital at Shakopee refused to treat Indians, and he was sent many hours away to the Pipestone Indian Hospital. His son Albert Jr. arrived a week later to find his father still covered with dirt and blood. A month after his accident, Albert Sr. died in the hospital of a fractured skull. The family shipped his body home on a train and buried him in St. Peter's cemetery at Mendota. They abandoned the land at Shakopee, and it lay fallow for years.

Lillian LeClaire and her children moved into the white world and didn't look back. Being Indian had brought them nothing but grief. As far as they could see, the same could be said for everyone they knew.

And it remained true until the LeClaire family was long gone from the reservation.

For years after the LeClaires left, the handful of Dakota families that still struggled to farm the Prior Lake Indian Settlement lived in poverty. In 1969 they formed their own tribe, the Shakopee Mdewakanton Dakota.

A decade later, a Florida tribe opened the nation's first Indian gaming operation, and the Shakopee Mdewakanton Dakota's tribal chairman, Norman Crooks, decided to take a trip to see it for himself. He was thinking about the possibilities for his own tribe, whose reservation sat just minutes from Minneapolis.

The fruit of Crooks's vision, the Little Six bingo parlor, opened in 1982 and paid off its million-dollar construction debt in six months. Mystic Lake Casino, added a decade later, became one of the most profitable Indian casinos in the nation. Three miles from the land Albert LeClaire once farmed, casino spotlights that were visible thirty miles away pierced the night sky, forming the shape of a Dakota tipi. The tribe's enrolled members—people who could trace their lineage back to the families who first farmed the land—were suddenly rich beyond imagining.

One Sunday morning, the first of May in 1994, Bob Brown opened the Minneapolis *Star Tribune* and an article headlined "Just Average People" caught his eye. The photo showed a woman named Cathy Crooks standing with her boyfriend in front of their trailer home on the Shakopee Dakota Reservation. Beside the trailer a Cadillac gleamed; behind it, the photo showed the shell of a new house. It was to be Cathy Crooks's dream home—three levels with a spiral staircase, slate and oak floors, two fireplaces, and a huge master bedroom with French doors—made possible by her share of the profits from Mystic Lake Casino. Cathy Crooks, the niece of tribal chairman Stanley Crooks, had just, in January, been accepted as an official member of the Shakopee Mdewakanton Dakota Community, a turn of events that qualified her for per capita payments estimated at half a million dollars a year. The article noted that the tribe numbered only about 150 people. Membership was based on degree of Mdewakanton blood and long-standing family ties.

Bob put down the paper and called Mickey. "Isn't that where Grandpa had his farm?" he asked. "She said, 'Absolutely, those Crooks are our cousins,'" Bob said later. What followed, he recalled in his usual understated way, was "kind of a movement in my family to get enrolled and get involved in all those riches they have out there."

Two weeks later, more than two dozen members of Lillian and Albert LeClaire's extended family gathered at Bob's sister Beverly's home. They had copies of Albert and Lillian's birth and death certificates. Family members brought copies of their own birth and baptism certificates and those of their parents. Then everyone filled out applications for membership in the Shakopee Mdewakanton Dakota. "I thought we were shoo-ins," Bob Brown said. "We thought we'd be welcome there."

The rejection letter came a year and a half later. It stated that the family did not meet the tribe's "descendency requirements." Still, the family knew they met the tribe's requirement of one-quarter Indian blood. They filed a lengthy appeal that included Albert's application for his farm, his name on Mdewakanton Sioux census and annuity payment records, and a genealogy showing that Lillian Felix LeClaire's brother, Peter Felix, was the great-grandfather of tribal chairman Stanley Crooks.

In February 1996, the family learned that their appeal had been rejected. What it came down to, they realized, was not a matter of how much blood they had. It was that Albert and Lillian's children hadn't stuck it out on the reservation. Now the families that had were reaping their reward. Tribal sovereignty gave them the absolute right to decide who got to share in it and who didn't, and Bob tried not to begrudge them their luck. "I know every one of them was dirt poor," he said.

But something did come of the LeClaire family's attempt to join the tribe. "The enrollment officer just offhandedly said to us, 'Go form your own community,' and in the end, that's what happened," Bob explained later. After the LeClaires' rejection, Bob had written an angry letter to the newspaper *Indian Country Today*, asking for other rejected families to contact him. Now the whole group, some 280 people, including distant relatives the LeClaires had met in the course of their research, decided to apply for federal recognition as a tribe.

"This was all about greed to begin with, I'll be honest with you," Bob Brown's wife, Linda, admitted.

At the time, Linda Brown cleaned houses to earn extra income for the family. She had shoulder-length, graying hair and dangly beaded earrings, a deep, smoker's voice, and an open, trusting manner. "In the beginning it was so we could get a casino. Why shouldn't we be able to? But as time progressed and we did more research and got more people involved, the focus shifted." The casino receded into the background. Linda Brown, who is of European descent, was struck by "how sad it was that so many of these people didn't have a clue about their Dakota heritage."

For Bob Brown, the most important thing to come out of it all was learning more about his own background, especially the story of his grandfather's death. "The absolute unfairness of it. What kind of arrogance would lead someone in an emergency room to turn somebody away?" he asked. "His hip was protruding out of his side, he had a massive skull fracture—it pissed me off. I decided at the time I heard that, that I was Dakota and I would be Dakota for the rest of my life." Bob Brown had always been interested in history, but now learning about the past made him angry. "The more you read about what happened, the more indignant you get. . . . When you think of your grandmother and grandfather and what happened to them, you just can't deny them, you can't deny your heritage."

Bob Brown never tried to hide the less-than-glorious path to his newfound identity. "There's another part of this," he would say. "Regrettably, this all started out over money." But, like Jim Anderson, he believed people can choose who they want to be: "We don't have to be Indian people but we *are* Indian people."

Still, it was easy for outsiders to be cynical. The Mendota Dakota had come together in the first place to try for a piece of a casino, and if they won recognition as a tribe, that might make them eligible for their own casino. For all anyone knew, that was all they really cared about. They could call themselves Dakota, they could take Dakota-language classes from Chris Leith at the Mendota VFW, they could go to Sundances and sweat lodges, but always the question of their real motives hung in the air.

That's how things stood in the summer of 1998, when Bob and Linda Brown got a call from Bob Greenberg, the Earth First! organizer.

Bob Brown learned later that before Greenberg called them, he had gone through a long list of established Native American leaders, but they had been wary of environmentalists with a sudden interest in Native sacred sites. Indian people had been used by environmentalists too many times, they said. Finally, Greenberg called Chris Leith, and Leith referred him to Bob and Linda Brown.

On a Sunday morning in July, Greenberg pulled up in front of the Browns' modest ranch-style duplex in Champlin, a middle-class suburb north of Minneapolis. At their kitchen table, he talked to them for three hours, telling them about Carol's house, the trees, and the spring. Since they were newcomers to Indian politics, the question of whether they were being used didn't occur to the Browns. "My first reaction was this was too close to Mendota, this was indeed part of our ancestral homeland, and I needed to get down there and see what was going on," Bob Brown said later.

A couple of days afterward, Bob and Linda Brown, together with Bob's nephew Michael and sister Bev, stopped by Carol Kratz's house. "She was beside herself, she was so happy," Linda Brown said later. Carol's friend Mary Jo Iverson, another local activist, gave the group a tour of the spring, the river, and the trees. "We knew then that it was something we had to do," Linda Brown said.

"We came there and very quickly learned what these places meant in the past and what they should mean to us—a lot that we didn't know," Bob Brown remembered. Someone pointed out that the four old oaks had multiple trunks, seven in all, one for each of the seven rituals of the Sundance.

"Seven sacred rites, seven trunks on those four trees," Bob said. "It just isn't happenstance. The first time I was told they

Bob Brown, chairman of the Mendota Mdewakanton Dakota Community, and his wife, Linda Brown. Photograph copyright Keri Pickett.

were put there for ceremonial reasons, I believed it." Bob thought there was a reason the highway had been delayed so long and the trees left standing. "I looked at it as they were self-protecting through all this time and now it was time for us to protect them. I've come to believe that all of this was orchestrated ... this was meant to happen."

At first, they couldn't understand why Chris Leith had given their names to Greenberg, but Bob Brown realized there had been a reason for that, too. "We are the Mendota people. We are the people that should protect it. I think Chris knew it and that's why he sent Bob to us."

In the fall of 1998, a Lakota elder, Harry Charger, visited camp from the Cheyenne River Indian Reservation in South Dakota. He reported seeing in the branches of the four oaks the spirits of children who had died at Fort Snelling. "Out of respect for our ancestors, I have to believe those trees were put there for a reason," Bob said later. "Maybe so we don't forget what happened here."

It was all part of a larger design, Bob believed. "We were destined to find out these things ourselves as part of the whole process of learning and being Dakota," he said. That's why they had been drawn to the protest camp in the first place.

"It really was an amazing change that came over all of us," Bob said. "Jimmy and Michael, they just moved in over there [at the protest camp]. They barely went home to clean up." Bob never heard voices himself—"I haven't been blessed with that connection, though I pray for it. I don't see things in the sweat lodge that other people do"—but the voices that others heard helped convince him. "I'm not a particularly spiritual man and I don't pretend to be—but to see the change that's come over Jim, it's incredible," he said.

Bob Brown made a wooden sign, painted it by hand in neat, precise letters, and took it to the protest camp. The words were taken from a book of quotations from Joseph Nicollet, the French cartographer who mapped the upper Mississippi and Minnesota Rivers in the 1830s. "The Mdewakanton people have always considered the mouth of the Minnesota River to be the middle of all things," the sign said, "the exact center of the earth."

Now, near the mouth of the Minnesota River, block after city block encrusted the bluffs. Freeway bridges spanned the steep green banks. Bob Brown and his family wanted to save a field, four bur oaks, and a spring. "It started out on principle," he said. "You've taken enough. You don't need your damn road here."

CHAPTER 5

"I'll Do Anything"

In her twenty years in the Minnesota legislature, Representative Karen Clark had always considered it an honor to represent Native Americans. Her district included an urban Indian neighborhood, one of the oldest and most concentrated in the nation, clustered along East Franklin Avenue just south of the skyscrapers of downtown Minneapolis. For more than thirty years, she had maintained friendships with leaders of the American Indian Movement, which began on Franklin Avenue in 1968.

The American Indian Movement started, wrote cofounder Dennis Banks, when "Native Warriors came together from the streets, prisons, jails, and the urban ghettos of Minneapolis" in response to "slum housing conditions; the highest unemployment rate in the whole of this country; police brutality against our elders, women, and children." Karen Clark, then a young nurse at the Hennepin County hospital in Minneapolis, volunteered to ride along as a medical observer as AIM members patrolled the streets. "I just provided what was called medical presence. I'd started doing that during the antiwar movement," Clark said. The AIM patrols picked up intoxicated men along Franklin Avenue and took them home so they wouldn't be arrested by the police and taken to the Hennepin County jail, where "they suffered a lot of abuse," Clark recalled. When she ran for election, AIM members were among her most loyal supporters.

Now a tall, slim woman in her early fifties, Karen Clark dressed in tasteful suits and slacks and mingled easily with other legislators in the polished halls and high-ceilinged rooms at the State Capitol, but she never strayed far from her activist roots.

In the mid 1990s, when the state's mighty utility, then known as Northern States Power, wanted to increase the amount of radioactive waste it could store at a nuclear plant on the Mississippi River southeast of Minneapolis, a stone's throw from the Prairie Island Dakota Reservation, Clark fought hard to stop it. When the legislature approved the storage, she was so distraught she considered suicide for the first and only time in her life.

"That was incredible. There is no community other than the Indian community that would have this forced on them," Clark said. "I've got to tell you, I've been a legislator for twenty years. There is racism against Indian people very deeply ingrained. It's just right there, right under the surface, all the time.

"You know, I feel like I have such a different viewpoint of Indian people. I mean, I am so enriched by living next door to them and representing them. I feel recently like they helped save my life. I had cancer last year, and part of what got me through it was some healing that happened with Indian medicine."

In January 1999, a few weeks after the December raid on the protest camp, Clark got a call from Sharon Day, an Ojibwe woman who headed the Minnesota American Indian AIDS Task Force. Clark regarded Day as a "deeply spiritual person"; the two women had worked together and been friends for twenty-five years. Now Day had a favor to ask. "Sharon said, 'I need you to understand what this means to our people. I want you to come with me to the spring,'" Clark recalled. "And she said, 'I also want you to come to the encampment where the four sacred trees are. I think if you just stand among those four trees you'll understand what you need to know.'"

The road to the spring goes through a steel gate, open on weekdays between 6:15 and 3:00. A sign nearby reads "This Federal facility (formerly the U.S. Bureau of Mines–Twin Cities Research Center) is closed. U.S. Property. No Trespassing." The cracked asphalt road winds through a parklike expanse dotted with shade trees and the hulks of abandoned buildings. Rust stains streak the 1950s aqua tiling on the building's exterior, the once-trendy salmon-pink steel doors, and the faded plywood that covers the broken windows. Signs saying "Danger, liquid oxygen" and roofs bristling with exhaust pipes hint at what once went on inside. On the main building, by the entrance, an eagle adorns a seal reading "E pluribus unum" above, and "safety, efficiency, mineral industries" below.

Past the entrance, the road goes over a small rise, and in wintertime, on the otherwise lifeless grounds, a loud quacking fills the air—hundreds and hundreds of ducks call and splash in a deep pool by a bare weeping willow. Springwater cascades into the pool from a limestone ledge.

The pursuit of safety, efficiency, and mineral industries is only the latest use to which this land has been devoted, as a nearby plaque explains. Erected by the Minnesota Historical Society, the plaque bears a golden silhouette of a settler and oxcart over the words "Camp Coldwater."

On May 5 of 1820 Lieutenant Colonel Henry Leavenworth moved the 5th U.S Infantry troops under his command to this area to escape the unhealthy conditions they had endured at their earlier stockade on the Minnesota River. The clear, cold spring water helped restore the men and their families, who lived in tents and elm bark huts here during three summers while they built the permanent stone fort nearby. The military continued to use the spring's fresh water through the nineteenth century, using horse-drawn water wagons and later a stone water tower and underground pipes to transport the water to Fort Snelling.

Families who left the Red River colony of Lord Selkirk were allowed by Colonel Josiah Snelling to settle near this location in 1821. Here they raised cattle and sold provisions to the army. When they were forced to vacate the military reservation in 1840, they moved downriver and helped establish St. Paul.

Blacksmith shops, stables, trading posts such as R. F. Baker's substantial stone warehouse, the St. Louis hotel, and steamboat landing all occupied this area, but by the time of the Civil War nearly all were gone. Today, this spring is all that remains of Camp Coldwater.

Hidden since the 1950s behind a fence on the campus of the former Federal Bureau of Mines in south Minneapolis, the Camp Coldwater Spring flows near the Highway 55 reroute. Photograph copyright 2001 *Star Tribune/ Minneapolis–St. Paul.*

Hidden behind the fence, the Coldwater spring had flowed unnoticed ever since the 1950s, when the Bureau of Mines first began building its campus. After the federal government shut down the Bureau of Mines in 1996, almost the only people who ever saw the spring were the skeleton crew assigned to keep watch over the abandoned buildings. Sharon Day, who lived just across the river in St. Paul, had never heard of the spring until the winter of 1998. But by then, word was beginning to spread.

In December 1998, a few weeks before the raid on the protest camp, Sharon's sister told her that Eddie Benton Banaise, leader of the Native American spiritual society to which they both belonged, was calling its members to the protest camp for ceremonies with the Mendota Mdewakanton Dakota.

Sharon Day went to a pipe ceremony at the camp that night. Banaise spoke of the spirits that still lingered near Fort Snelling, where the Dakota had been held captive so long ago. "He talked about the babies that died here that need to be remembered. He said we will help them but in turn they will help us," she said. Later that night, she went to a sweat lodge on the Mississippi River behind Carol Kratz's house. Both Native people and Earth First!ers attended. A full moon shone, and just downriver, some thirty feet away, they could see a group of witches, or modern-day followers of the earth-based religion known as Wicca, holding their own ceremony. "They were singing songs and chanting and we were in the lodge," Sharon Day said later. She felt there was a reason the two groups were in the same place. "You see, there is this metaphysical relation we have with the earth. And there are certain places of the earth, they call us. They don't just call, 'Hey, you Native Americans.' They call everybody."

Not many Native people used the spring before the Mendota Dakota came along, Sharon Day said later, "but now, we do. That's where we get water for our ceremonies, our sweats." She

and her sister went to the Coldwater spring and sang the traditional water songs of Ojibwe women. "I believe that nothing
that humans can do can hurt the spring. Those water spirits are
very strong, and what they needed was for us to remember them."

So in January 1999, Sharon Day and Karen Clark went to
the Bureau of Mines campus and stood by the pool and the
Coldwater spring and the big weeping willow. "We didn't talk
much," Sharon Day said. "I just wanted her to feel those water
spirits at the spring. She had tears streaming down her face. There
wasn't much need for words."

The two women walked out the gate and across the snow-
covered field to the four oaks and put cedar and tobacco on the
fire beneath them. Karen Clark touched her hand to the trees.
"The reverence people had there was very present," Clark said
later. "It was a very powerful experience. I don't know what to
say other than that."

It seemed to her that there must be another route for the
highway. In addition to her revelation at the trees, she had other,
down-to-earth reasons to try to find an alternative. Although the
planned route lay just outside the official boundaries of Minnehaha Park, it went through green lawns and trees bordering
the park—land that many of her constituents had always thought
was part of the park. "I was getting calls from all kinds of people,
saying, 'What in the world? Is the city allowing a freeway to go
through a park?'" she said. A senior citizen group learned that it
couldn't hold its annual picnic that year because of the planned
construction and called her in dismay. "People just couldn't understand it."

Clark arranged meetings between the Mendota Dakota and
officials at the Department of Transportation. She also introduced
a bill to recognize the Coldwater spring as a "traditional cultural
property which is entitled to preservation and protection."

The argument that it was too late to change the road's planned route didn't faze Clark. "I can't tell you how many times that sort of attitude has been there for everything," she said. She did a mocking imitation of the critics: "You should have come to us with that long ago, because now you're going to jeopardize federal funding, you terrible people. How dare you jeopardize federal funding to do this very needed transportation system?" Clark thought there was more wiggle room than the Department of Transportation wanted to admit, and she had seen deadlines changed before.

With Karen Clark on their side, it seemed to road opponents that working within the system might work. They had a bill before the legislature. Then, in February 1999, it suddenly seemed they had a chance in the courts as well.

Just after the raid, Jim Anderson and Bob Greenberg had stayed up all night working with lawyers to get a lawsuit against the reroute filed by Christmas Eve. Now that lawsuit had borne fruit. On February 2, a Hennepin County district judge ordered federally supervised mediation between the Department of Transportation and highway opponents. At the camp, Bear handed out flyers. "STRATEGY MEETING FOR MEDIATION" they read. "INDIVIDUALS OR REPS FROM ALL COALITION GROUPS WELCOME!!!"

On Sunday afternoon, February 7, thirty people met in the basement of All Nations Indian Church, a cement-block building just south of Franklin Avenue. They burned sage and sweet grass, then went around the circle introducing themselves. Twin Citians sometimes wondered who, exactly, were the "protesters" they saw on the news. The basement meeting was a fairly representative sampling.

Bob Brown stood up in his red "Mendota Mdewakanton Dakota Community" warm-up jacket, along with other members of the LeClaire clan and their relatives. Caleb and Tarzana, the two who were among the camp's hardiest winter residents, had come for the meeting. "I've been a happy camper since Halloween," Tarzana joked.

A young woman named Emily said she was doing some legal work for the cause. A graduate of Macalester College, a private liberal arts school in St. Paul, she still looked like a college girl, with short, shiny brown hair and the air of a good student.

"I'm Botany Bob," said a man who called himself an "eco-warrior" and "Earth First! hellraiser." "They're going to bulldoze those oak trees over my dead body," he announced.

A middle-aged Native American man who introduced himself as Larry said he had lived at the camp for six months. "It's like a family to me," he said.

A few members of the neighborhood groups who had fought for years against the highway had come as well. They tended to be people who believed in working through the system, not chaining their arms to cement. They worked hard behind the scenes and got relatively little attention. "They all looked kind of the same to me," Natalia admitted much later. "I mean, like, they were all kind of grown-ups. They were the Sierra Club and Green Cities and all those people. They really weren't the same, and I know that now, but back then they were just sort of, like, the coalition. And they all went to meetings, which we really didn't go to at that point."

One more ingredient in the mix was the all-purpose activists, the people whose names appear time and again as organizers for events to free Mumia Abu-Jamal or support the Zapatistas. Some were mellow, sixties types. Others were strident, with a bitter edge, as if they had been ranting for years to no avail.

Nobody seemed to be in charge of the strategy meeting, and although the flyer had urged people to bring "clear and concise written ideas and concerns," no one seemed to have done so. But what the meeting lacked in strategy, it soon made up in rhetoric.

The main speaker was Michael Haney, a flamboyant, charismatic Seminole from Oklahoma who wore his graying hair in thin braids down his chest and gave his title as executive director of the American Indian Arbitration Institute. He breezed in more than in an hour late, sporting a jacket with Seminole designs in colors as bright as his mood. "No one ever envisioned that we would have the state ordered into federal mediation, discussing federal laws. That's a tremendous coup," he announced. Haney predicted that Karen Clark's legislation would, for the first time ever, extend equal protection of religious freedom to Indians in Minnesota, the way the American Indian Religious Freedom Act had nationwide. "We're asking that our sacred sites have the same standing as their churches," he said. "If you burn down one of *their* churches, they'll hang you and stuff you and put you in one of their museums."

Michael Haney had been meeting with members of the Meskwaki tribe in Iowa, he informed the group; the Meskwaki had expressed "tremendous support." Haney said many tribes recognized the sacredness of the Coldwater spring, and they could invoke federal laws to protect it. Tribes had been able to prove their claims about sacred places in court before, he said, and they could this time, too. "We've lived here fifteen thousand years," he said. The meeting went on for hours before it finally broke up. Bob and Linda Brown listened, their hopes rising.

They and the other Mendota Dakota had faith in Michael Haney. They gave him a place to stay at their houses, fixed his car, and lent him money. Everything he said about legal avenues seemed promising. He also encouraged them to lobby at the State Capitol alongside other members of the coalition.

None of the Mendota Dakota had ever lobbied before—"It

was totally unexplored territory," Linda Brown said later—but they were willing to try. Before they made their first trip to the Capitol, Bob Greenberg held a training session, but once Linda was inside the echoing marble-floored halls, she felt too nervous to say much. "It was like, what the hell are we doing here?" she said. Bob's sister Linda M. Brown took the lead. The two Lindas and a handful of others got a directory and found their way from office to office, keeping a series of appointments that Greenberg and others made for them.

"I didn't think we had a snowball's chance," Linda Brown said. But as the days went by, her hopes rose. Karen Clark's bill passed unanimously through a House governmental operations committee. "Everyone started crying. We had to go out in the hall," Linda Brown said. The bill's Senate version picked up five sponsors. "Nobody thought we'd get a senator. We got five."

"You know, people were getting through, particularly the Indian women like the Lindas, they were making inroads," Karen Clark said. "They spoke with their hearts and minds in a way that really started connecting with legislators."

On a warm, bright winter weekend at camp, a big pot of vegetable soup—the raw materials scrounged from Dumpsters—simmered on a gas burner in the makeshift kitchen. Nearby, several Earth First!ers, including two called Moon and Spiney, sat around the campfire, deciding who would fetch water from the Coldwater spring. Spiney was six feet tall, with a combat jacket, heavy black lace-up boots, and a silver nose ring dripping in the cold. A plastic test tube hung from one pierced ear, reaching nearly to her shoulder.

Nearby, Jim Anderson and a cluster of Indian people emerged from a large, brightly painted tipi underneath the four trees, where they had held a ceremony. The tipi was packed, Jim said happily to a visitor. The two groups, Indians and Earth First!ers, seemed to exist in two separate spaces. The Earth First!ers were there to support the Native Americans, but some among the Mendota Dakota wondered whether the young protesters, well meaning as they were, weren't more of a hindrance than a help.

When she first met Spiney, Linda Brown was taken aback. "That Spiney, she scares me. I'm sorry," Linda said later. "The first time I saw her she had that eyedropper thing in her ear. I thought, *what* is going on here?" Linda thought of herself as a conventional, middle-aged person who had never protested anything in her life. "I'm a *grandmother*," she said. Linda and other Mendota Dakota worried that the field was starting to look like a camp set up by homeless people. They knew of Indian elders who visited once and never returned.

The new camp sat on sacred ground, the Mendota Dakota felt. They didn't call it Minnehaha Free State; they called it the Minnehaha Spiritual Encampment. Advised by Chris Leith and another elder, Harry Charger of the Cheyenne River Sioux Reservation in South Dakota, the Mendota Dakota believed they could stop the highway using prayer and legal avenues, not Earth First! tactics. They told the Earth First!ers to take the chains off the four oaks.

Jim Anderson knew that a lot of his relatives wanted Earth First! to leave the camp. "They said a lot of elders didn't like coming down there because people were dirty and all this stuff," Jim said. "Well, some of 'em *were* dirty, you know, and it's hard for me to judge somebody because they don't want to wash and use chemicals that are polluting the earth. I mean, that's why they don't use soap and water, I guess." But *somebody* needed to be

there to hold off MnDOT, he insisted, and it wasn't easy finding people willing to sleep outside in a Minnesota winter. "It was
damn cold. It got pretty lean out there."

Jim knew there were differences between the Indians and
the Earth First!ers. The Indians ate meat and believed in leaders.
Many of the Earth First!ers were vegetarians or vegans and
believed in deciding things by group consensus, which often
meant sitting down in a circle for hours and holding seemingly
endless earnest discussions. But Jim was friends with the Earth
First!ers. He couldn't just tell them to leave. "These people have
been loyal to me, and I knew they were all there in a good way,"
he said. So the Earth First!ers stayed.

For a while, people slept on a platform sheltered by a wall of
plywood. Then somebody got the idea of building a dormitory.
Out of pallets and blue plastic tarpaulins, they created sleeping
pods radiating from a common center. They called it the Starlodge but soon discovered the drawback to its fanciful floor plan:
it was impossible to keep warm. It shed heat as efficiently as a
car radiator.

Dr. Toxic and Natalia, the feisty animal rights activist
couple who had locked down together in the basement during the
raid, moved back to the camp in late winter, a few days after they
hung an animal rights banner at the University of Minnesota.

"I got trained to climb, then I went up to the top floor of
Moos Tower, which is a twenty-some-odd-story building," Dr.
Toxic explained. "I went out on the roof and dropped a rope,
and Natalia stayed at the top."

"I was his support person. I literally pushed him over the
edge," Natalia joked.

The banner said "Stop Primate Research." Dr. Toxic rap-
pelled down the tower's face and dangled in midair for five or
six hours before police pulled him in through a window. He had

wanted to stay up much longer, but being near the window had been his big mistake, he realized. The next year, a guy dropped a banner near the tower's corner, where there weren't any windows, and *he* held out for days. It was something to remember, Dr. Toxic figured, for the next time he tried something like that.

At camp, Dr. Toxic and Natalia helped build "tree sits," or stations with platforms or sleeping hammocks where activists could hold out in trees. They helped organize food donations, carried water from the Coldwater spring, and did dishes, a chore that under the primitive conditions at the camp sometimes took all day. "We'd spend our days just hanging out in the woods and doing whatever work needed to be done," recalled Dr. Toxic.

Tumbleweed, who had been so excited about the first camp, never lived at the second one. As a free-spirited Earth First!er and self-described anarchist, he found it hard to put up with all the rules. Compared with the Free State, the new camp had lost

Dr. Toxic in a tree sit.
Photograph by Peter Hoseth.

its edge, he said, what with all this emphasis on working with the system instead of direct action. Plus, spiritual advisers who didn't even live there were making decisions for the camp. "It became one of those things," Tumbleweed said, "where people were trying to control—you know, just saying, 'Oh, this is a spiritual encampment, you're not here to, like, have sex.' People were hooking up, and people were getting to know each other and people were falling in love, and people were told, 'No, you're not here for that, and if you want to support this, you have to do it our way.' I respect their spirituality, but I wasn't there to defend a sacred site. I was there to resist global capitalism in my backyard."

Like a number of the Earth First!ers, he seemed to prefer direct action campaigns to sitting through the tedious meetings necessary for going through the legal system, but he still stood behind the Mendota Dakota and their cause. That winter he worked at a recycling center but kept in touch with how the campaign was going.

At camp, the Earth First!ers waited. Now was not the time for direct action. The construction season was still months away. As the winter wore on, people put a layer of wood chips around the fireplace and the kitchen and the four bur oaks to keep the soil from being trampled into mud by visitors. On weekends and evenings, the campers and their supporters held celebrations and ceremonies. On warm days, the carpet of decaying wood chips filled the air with the rich, sweet smell of a forest floor.

On weekdays, the camp was often nearly deserted, and the motley collection of battered lawn chairs and uprooted bus seats around the campfire sat empty. Solitary in the field, Bear would sort through piles of donated clothing. He folded greasy-looking down jackets into garbage bags to take to the Free Store in Minneapolis. He made sure the ceremonial fire under the four trees was kept burning, day and night.

As Bear and the Earth First!ers kept watch out at camp, Jim Anderson's conviction that it stood on sacred ground grew. Someone gave him a 1976 document from the Minnesota Historical Society's Historic Sites and Archaeology division that listed a "Dakota Burial Ground" near Fort Snelling and cited observations made in the 1860s by the missionary Gideon Pond. "A little to the left of the road leading from Fort Snelling to Minnehaha, in sight of the fort is a hill which is used, at present, as a burial place. This hill is known as 'Taku wakan tipi,' the dwelling place of the gods. It is believed that one of this family of divinity, the Onktehi, dwells there," Pond wrote. Whoever had filled out the form for the Historical Society in 1976 had noted that the "exact location and present condition" of the burial site were unknown.

Jim had another citation, from the newspaper the *Dakota Friend*, published by the Dakota Mission under Pond's editorship, describing the same area in more detail in 1852:

> *Immediately above the confluence of the Mississippi and Minnesota, of St. Peters rivers, is spread out a most beautiful prairie ... bordered by a belt of oaks, which in the distance rest and delight the eye, especially when verdant with their summer dress. Fort Snelling is built on a sharp angle of this plain pinched up between the thumb and finger of the Mississippi and Minnesota. About one mile west from the fort, starting from a point near the river and extending southwest through the prairie is a bluff or ridge which by Americans is called Morgan's bluff, and by Dakotas, Taku wakan teepee, the dwelling place of the gods. On the top of this bluff, in sight of the fort is a little cluster of children's graves who have died at or near the garrison. The Dakotas think that one of their superior gods live here under the bluff, and believe that he has often been seen by some of their people. They call his name Oan-ktay-hee. Being an inhabitant of the water, and the earth deep under the water, he will answer to the Neptune of ancient heathen. He is the god of medicine, and the celebrated*

medicine dance is made in honor of him, and the songs that are sung on such occasions are those which the medicine men have learned from the Oan-ktay-hee.

What he saw in the archives just confirmed what Jim already believed. "Everything I've been saying is history, it's all true. I haven't had to change anything," he would say. He knew that Morgan's Mound was the name white settlers gave the hill where the Veterans Administration hospital now stood, right there in south Minneapolis. You could stand on its lawns and look down the slope and there were the four bur oak trees and the camp. But the authorities and the public didn't seem to listen when he talked about Taku wakan tipi and the water god, Oan-ktay-hee. "They don't want to deal with history," he said.

Still, Jim kept looking for the historical proof he hoped would somehow save the land. That winter, he and his uncle Leo and a couple of history buffs who opposed the highway went to the Minnesota Historical Society to see Allan Woolworth, Minnesota's former state archaeologist. Woolworth was retired but still went daily to his office in the society's headquarters, an imposing mass of layered stone that looked like a cross between a fortress and a torte. As an institution, the Historical Society had kept its head low during the dispute over the highway.

Woolworth was an expert on Dakota history, coauthor of a book on Little Crow and coeditor of *Through Dakota Eyes*, a collection of firsthand Indian accounts of the Dakota conflict. He himself had been a bit player in a long-delayed epilogue to that conflict. In 1971, at the request of Little Crow's grandson, Woolworth took Little Crow's remains from the Minnesota Historical Society shelf where they had been stored since the early 1900s and carried them to South Dakota, where they were buried in a private family plot.

Woolworth, who described himself as the society's "resident heretic," told Jim and his friends there was a good chance there were graves in the area, of both Dakota and early settlers. He advised them to delve into the local military history and push to have Camp Coldwater put on the National Register of Historic Places. He suggested they apply to Washington, where he thought they would have a fairer hearing than in Minnesota. "Write that down," said Jim. "I'll do anything."

Even MnDOT's Bob McFarlin believed that Jim Anderson was sincere. Although he didn't agree with him on the issues, McFarlin respected Jim's passion. He felt that Jim was honest, although on the wrong track. "Jim would in meetings start talking about all of the history of the Mendota, and I'd say, 'Jim, I have no reason to doubt you, you may absolutely be right, but we're not the ones to convince,'" McFarlin said.

To McFarlin, whether or not the road was built boiled down to a simple matter of process. The Department of Transportation had to comply with myriad regulations, and as long as it did, the road would go through. And the department had jumped through every hoop. It had gone through the proper procedures with the State Historic Preservation Office and the Federal Highway Administration. The Minnesota Indian Affairs Council, a state agency made up of the tribal chairs of Minnesota's eleven federally recognized tribes, had not raised objections to the road. MnDOT spoke with the council frequently, McFarlin said. "The council wanted really nothing to do with any of it publicly. In our meetings privately they were not supportive of Jim Anderson's claims. They didn't offer any thought that that was a site that needed protection.... They were very worried about association with members of AIM, that were at times at the encampment, and the folks from Earth First!."

So when the judge ordered mediation between MnDOT

and highway opponents, Bob McFarlin was angry. The law laid out a process; the department had complied. "What was there to mediate?" he asked. McFarlin and other MnDOT officials met with lawyers from the state attorney general's office at MnDOT's office building, eight stories of unadorned granite, the tallest building at the State Capitol complex. As McFarlin saw it, the officials had no legal room for maneuvering. They couldn't agree to anything outside the process. "There wasn't at that point any area of discretion or compromise that we could even talk about," he said. But as it turned out, Michael Haney, the flamboyant speaker from the American Indian Arbitration Institute, did most of the talking at the mediation. He wore a dark suit and wingtips polished to a high gloss.

McFarlin didn't trust Haney. "He just sort of plopped into the scene from Oklahoma," McFarlin said. "We didn't know who he was, and nobody else seemed to know who he was." Haney made a lot of claims: he had represented this tribe in Iowa, or that tribe in Kansas; he had high-level contacts in Washington. But as least as far as MnDOT could ascertain, Haney's claims didn't check out. "There was no paper trail of all the things he claimed he had done. People knew of him, but it was very nondescript." MnDOT had the state attorney general's office run a criminal background check on Haney, but it turned up nothing. Word from the investigators was that Haney was harmless, but not as important as he pretended to be.

"It was really interesting because he struck such a majestic type of presence," McFarlin said. "He could speak so well and so seemingly authoritatively about federal law. Legislators were very impressed with him. But you got beyond five minutes, and you realized there wasn't much depth there. This was a person who could talk for hours with five minutes of material."

At the mediation, McFarlin and other DOT officials listened to Haney for hour after hour. McFarlin was surprised at how much deference Jim Anderson, Bob and Linda Brown, and the other Mendota Dakota paid to Haney. "He was the negotiator, and everybody else sat and listened," McFarlin said. Haney even squelched the normally voluble Bob Greenberg, the prolific writer of Earth First! press releases, when he tried to talk. "Haney was in charge," McFarlin said.

At the end of the session, McFarlin and the lawyers and officials went to a separate room and conferred. They were prepared to offer what in McFarlin's opinion amounted to just about nothing. They would have a consultant do a study to determine whether the site the highway would pass through was sacred. Before the study was released, the department would give the Mendota Dakota a chance to review but not edit it. MnDOT had been planning to do the study anyway, McFarlin said, and had already chosen the consulting firm, New Jersey based Louis Berger & Associates, Inc. (now called The Louis Berger Group Inc.).

The Louis Berger Group's main business is designing and building transportation projects such as highways, bridges, and air bases. Since its founding in the 1950s, when it designed New Jersey's first interstate highway, the company has helped plan, design, or manage the construction of more than 100,000 miles of highway across the world, including the two-thousand-mile Trans-Amazon Highway that opened the Amazon Basin rainforest to development. Berger has in-house archaeologists and historians to do the cultural studies that may be legally required before a project can be built. It was archaeologists from this staff, known as Berger's Cultural Resource Group, that MnDOT planned to hire to do the Highway 55 study. It was not, McFarlin thought, a major concession.

"Really, really minor stuff," was how McFarlin later summed up what MnDOT and its lawyers were willing to put on the table. "We expected to go into the next phase of the meeting and say, 'This is all we can do.' We really expected the mediation to break down."

The officials and lawyers returned to the meeting room and made their offer. "The gentleman from Oklahoma bolted up out of his chair, stuck his hand out to me across the table, congratulated me on meeting their demands and reaching an agreement, and said what a historic day it was," McFarlin said. The road opponents filed out of the room, leaving the surprised MnDOT officials and their lawyers behind. "We met afterwards and kind of looked at each other and said, 'Huh?'"

Winter in Minnesota means day after day of cement-colored skies and dull light. But Sunday, March 14, was the kind of bright, blue-sky day when the world becomes multicolored again. In the field near the four oaks, the hues of the coats, caps, and mittens of hundreds of people shone brightly against the snow. A six-piece brass band, complete with tuba, oompahed cheerfully. A neighborhood activist known as Dan the Oak Man Keiser directed more than eight hundred people to their places. Like a marching band on a football field, everyone stood in formation, making letters that said STOP 55! GO OAKS!

The crowd included not just Mendota Dakota in red warm-up jackets and Earth First!ers in combat boots, but also the kinds of families you would expect to see at a Minnesota Twins game: dads and moms in nylon parkas, kids in snowmobile suits. Mingling among them were Karen Clark and a handful of other state legislators, including Gary Laidig, a state senator from the Mississippi River town of Stillwater. Laidig, a Republican, seemed an unlikely ally, but he had fought MnDOT before.

Years earlier, he had helped thwart the department's plans when it tried to condemn a wide swath of land in his district on either side of Highway 94. "It was brutal, bloody, and I'm still paying the price," he said. "Big labor, big business, everybody gets together on these deals." Now he hoped the state legislature would intervene in the Highway 55 dispute. Earlier, he had brought a couple of his colleagues out to see for themselves the urban greenspace that would be lost. He knew the key would be to convince legislators that the highway's opponents were mostly ordinary people, not radicals. "Everyone thinks it's just a bunch of hippie protesters," Laidig said.

A small plane taking pictures buzzed across the blue sky, and everyone cheered. Michael Haney took up a bullhorn. "We're tired of being characterized as urban malcontents! I'm proud to be a tree hugger!" he yelled to the crowd. "You are the new Indian

Eight hundred people protest the reroute in a field near the four bur oaks, March 1999. Photograph courtesy of Dan the Oak Man Keiser.

people! We waited five hundred years for you to talk about Earth Day, but now you're here!" The crowd cheered, and some howled, like wolves lifting their noses to the moon: *ahhooooh!*

Less than a week later, near the weeping willow and the Coldwater spring on the federal Bureau of Mines campus, Eugene Begay, an Ojibwe elder from Wisconsin, held a ceremonial pipe to the sky, praying in Ojibwe. He voice rose and fell in repeated, rhythmic phrases: the chanting heard in places of worship around the world. In the circle of people around Begay, a small group of men and women stood out from the rest. They wore business suits and shiny shoes; they represented the Federal Highway Administration, MnDOT, and Louis Berger & Associates, Inc.

The Berger consultants had come to gather evidence. They planned to weigh oral testimony from Indian people along with evidence from historical records, then decide whether any of the land affected by the highway project might be classified as a "traditional cultural property." Under federal law, such a designation would make the site eligible for protections designed to safeguard Native American religious freedom. From her place in the circle, Linda Brown noticed that when the ceremonial pipe was passed around after the sage was burned, almost every one of the officials and consultants put it to his or her lips. A good sign, she thought.

"I'd like to thank the entity that's responsible for all this, and that's the Creator," Jim Anderson said. "'Cause he pushes us in ways that we don't realize, sometimes. And look at the way he's pushed this." From a short distance away, a television crew filmed Anderson's speech for the evening news. At Begay's request, the crew had turned off their camera during the elder's prayer. "I never wanted to be in front of a camera," Jim said. "This isn't what it's about. It's about what this place is, what it means to our people." A small plane droned in the clear sky, the distant traffic sighed like the ocean, and a cardinal whistled from a nearby branch. Jim's voice wavered. "And it *is* a traditional cultural property. I've said that from the beginning, and I'll say that till I die. Please save it. That's all we're asking."

Sharon Day and her sister sang a ceremonial water song in Ojibwe, their strong, clear voices rising in unison, to the beat of a single rattle. They brought a copper kettle of water dipped that morning from the spring and passed it around the circle for people to drink. "Among our teachings, we Native women are responsible for taking care of the water," Sharon Day said. "We're the ones who collect it, we bring it to the ceremonies, we sing the water songs, we bless the water. We know that coming into the world, we live in water. That's the safest place that we've ever been.

"I was up in Winnepeg this winter and the grandmothers there, the teachers were talking to us about the water. And one of the grandmothers shared with us that we need to do more to take care of the water. This spring right here, this is the only spring that's left in the Twin Cities, and we need to take care of this.... This is such a beautiful place, and we have so few beautiful places left." She described the sun on the water, the happiness of mind and body she felt in such a place. "If the day ever comes when they bring the heavy equipment, I'll be here, and I hope you will be here with me." *Migwetch*, she said, the Ojibwe word for *thank you*.

Their hands folded in front of them, the officials and consultants listened gravely. The day before, at the State Capitol, they had heard eight hours of testimony from Indian elders. After the ceremonies were finished that morning, they would listen for another half a day. At the hearings, the words of the Indian elders seemed to come from a slower-paced world, from a time when people sat around campfires, listened to stories, and watched the constellations shift across the skies. For almost an entire morning in a crowded meeting room, Don Wanatee, a Meskwaki elder from Tama, Iowa, had meditated on a vast array of subjects and told stories from his life.

But occasionally Wanatee seemed to address the topic at hand. The confluence of the Mississippi and the Minnesota is one of a number of places where water spirits live, he told the officials. "The names that we use, of course, we have ten places," he said. "One is the headwaters of the Mississippi when we pray during this water ceremony, that's what we talk about. There's a spirit that resides there, watches over the headwaters of the Mississippi. That's in Itasca. We travel up there, offer tobacco once a year. Then we stop by here on our way up there.... There's two that sit here. There's one on this side and one on the opposite side of the Minnesota River where it enters the Mississippi, there's two there. Then there's one at the Wisconsin River where it enters the tributaries as a tributary of the Mississippi, there's one there. The next one is at the Des Moines River as it flows across Iowa from the state of Minnesota.... So we're all very aware of these rivers, we're very familiar because that is our world, our spiritual world. It's shown to us."

Chris Leith, the Dakota spiritual leader and Sundance chief who guided Jim Anderson and the other Mendota Dakota, held a plastic bottle of drinking water he had smuggled into the Capitol meeting room, where food and drink were prohibited. "I'm going to pray with this water. Water is medicine, very sacred," he said. He wore a baseball cap, flannel shirt, and jeans. He had deep sunken eyes, long gray hair, and a face that could have looked out from a century-old photograph. He spoke slowly, with a gentle Dakota accent. He knew some people thought the four trees were too young to have been used as burial scaffolds, but it didn't faze him. "That's their mentality," he said of the critics. "A tree doesn't have to be very big. Sometimes they grow slow, sometimes they grow fast. Depends on the weather, just like corn. You have to take these things into consideration. But they don't. Everything's under scientific terms."

Now, testifying before the highway officials and consultants, he tried to explain a way of thinking that was foreign to them. "My education comes from my grandparents of the spirit world. They're the ones that teach me, tell me," he said. "When I was out there at that site, we had a ceremony and we were told that years ago there was a Dakota village there. Some of them died there in burial grounds. Some of them suffered, and that's when the army held them prisoner and killed them, murdered them. The little children were used for target practice. The soldiers pretend to play with the children, tell them to run, and when they're running, they shoot them. Those are the kind of stories we used to live with.

"You don't find them in history books, but yet the village that was there, they were taking care of that sacred site. That's the reason they had the village there. My grandparents talked about a long time ago, all along the Minnesota River there is a village. They were taking care of that place. . . . We're here to take care of the way it was meant to be from the beginning of time. It is gone too long now, and now it's time for us to speak up."

Leith mentioned the four trees: "We have had ceremonies there with the Bigfoot Riders of South Dakota, who came and had ceremonies. . . . Those trees were scaffolds years ago when we were young. Some of the remains are still down below the roots of those trees. There are remains, skeletal remains," Leith said. "The question I have in mind, what happened to the remains of those that were murdered by the soldiers? There were thousands of people that were killed. We can't find them. What happened to them?" he asked.

Some of what Leith said, such as the story of the Dakota children used as target practice, has never been documented, but Leith told the officials that whether something was written down shouldn't be the test of whether it is true. "Our teaching is oral, it doesn't come from books. So when you talk about things like that, nobody believes us because it was not documented. It never happened, that's their belief. But that piece of paper can kill you, can put you to death. That has more power, it seems, than what comes out of our mouth."

Eddie Benton Banaise, the grand chief of the Three Fires Midewiwin Society, spoke also of oral tradition. "I was born into a full-blood Ojibwe Anishinabe family. Both my mother and father were full-bloods. My grandmothers and grandfathers as far back as we can search were full-blood. And so I am that today," he said. (The word *Anishinabe*, meaning "First People," applies to the ancestors of the Ojibwe, who migrated to the upper Great Lakes area from the Atlantic coast.) "I was born into a family that believed in the original religion, in the original spirituality of the Anishinabe people, the Ojibwe Anishinabe people, and I was raised in that tradition.

"The place in question I believe can be validated as being a special place, a sacred place, a sacred, cultural spiritual place. Through our oral tradition, our history recent and older, we know that the falls, the falls which we know as *Bawpi gaw ki ji waung*, which came to be known as Minnehaha Falls, that there was a sacred place, was a neutral place for many nations to come." The confluence of the two rivers was also a neutral place. "And somewhere between that point and the falls, there were sacred grounds that were again mutually held to be a sacred place. And that spring from which the sacred water should be drawn was not very far, and I've never heard any direction from which I could pinpoint, but that there's a spring near the lodge that all nations used to draw the sacred water for ceremonies."

Banaise said the Dakota, the Sac, the Fox, the Potawatomi, the Mdewakanton Dakota, and the Meskwaki people had all used the Coldwater spring for ceremonies and mutually agreed that it was a neutral and sacred place. "That is confirmed by our oral history," he said. "It is difficult to estimate when the last sacred ceremony was held intertribally, but my grandfather, who lived to be 108, died in 1942, and I will tell you this: Many times he retold how he and his family, he as a small boy traveled by foot, by horse, by canoe to this great place to where there would be these great religious spiritual events, and that they always camped between the falls and the sacred water place. Those are his words.

"I can't say any more than that because I do not know any more than that, " Banaise told the officials, but he noted that he had brought with him anthropological and archaeological studies documenting shared ceremonies by the Ojibwe, Dakota, and other tribes. He held up a book, by a Minnesota lay archaeologist, that described a common ceremonial ground. "And so I offer these things to you. Hopefully they'll be helpful."

Winona LaDuke, the Harvard-educated director of the White Earth Land Recovery Project on the White Earth Reservation in northern Minnesota and Ralph Nader's running mate in two presidential elections, was among the last to speak. She said federal laws designed to protect Native American religious freedom recognize that "certain lands and natural formations are inextricably intertwined with the practice of traditional Native American religions. For some tribal religions there may be no alternative places of worship. We are not relics in museums. We are live people who require these places to sustain ourselves so we can be who we are intended to be by the Creator. The reality is that there is very little left of these sites, you know as well as I do," she told the officials. "Antiquities and sacred sites are easily obliterated by concrete. They cannot be recovered."

LaDuke said it was a "strange irony" that one culture was given the power to decide for another what they held sacred. "I solely ask you to do the right thing and pray that you will hear these words that my elder spoke before," she said. "You know, you cannot have better people here. We are blessed every time we hear these people talk ourselves. And so I ask that you hear their words."

Eugene Begay, the Ojibwe elder from Wisconsin, prefaced his testimony with these words: "This morning in my hotel room I asked the Morning Spirits, I told them that I was going to use my best judgment on what to say about spiritual things, to let the Creator know that I'm doing this in a good way, that I have no personal gain for myself in declaring these things. And I also said even though I do not know the people I may be addressing, I said I pray that they be honest, respectful people who understand that what myself and others are going to be testifying here, are going to be saying, our spiritual values—those things that guide our life—are important to us."

Then Begay held up a copper rattle. "There was a noise that started it all and here is the noise," he said. Silence, then a sound like a rattlesnake. *Sheeeeeeeesh. Sheeeeeeeesh. Sheeeeeeeesh. Sheeeeeeeesh.* "The Ojibwe creation story takes fourteen days to relate its entirety in the Grand Medicine Lodge of my people, the Midewiwin. It takes fourteen days of singing, talking, and dancing and all the other rituals that are connected with it. But I'm going to tell you just excerpts of the creation story that I feel are pertinent to the land and the road and the water we're talking about here."

Begay told how the Creator of the world had a vision of the

stars, the earth, its animals and plants, and the original people, the Anishinabe. "Human beings were created lastly. We are the most dependent upon all the others.... We are weak and we depend upon the rest of God's creation. That is why you heard this morning among my predecessors here testifying the importance of the land, the water, the importance of our relationship one unto another."

Begay ended his testimony with an Ojibwe Medicine Lodge song, one of the few that can be sung in public. "It's a long song, but I'm going to abbreviate it for you. I'm only going to sing it four times through. Music is a form of spiritual communication. I asked my uncle one time, 'What does Heaven look like, what does the spirit world look like?' ... He said, 'I can't explain it to you.' He said, 'Give me that drum over there,' and I did, and he sang me a song. And when he got done, he said, 'That's how the spirit world looks, that's it.'" Begay shook his rattle again and sang to its steady, quick beat.

As they had for all the elders' testimony, the highway officials and consultants listened gravely, but asked no questions.

When the hearing was over, Linda Brown was sure MnDOT's consultants would realize that the land the road would cross was sacred. "How could they listen to all these people and not be convinced?" she asked. Every day she felt more optimistic. It was not just the elders' testimony, or that things seemed to be going well at the Capitol. It seemed to her there was a reason why in thirty years of trying, the highway had not been built. Something must be stopping it.

"Chris Leith said the ceremonies have been done, the spirits are there. They're protecting it. And so it's not going to happen, that's the way I feel," she said that March. "The spirits talk to him—I know this sounds strange—and the spirits have told him they're protecting it and the road isn't going to go through."

CHAPTER 6

Piestruck

Bob Greenberg bought the pie on a morning in late March, after reading an article in the March/April issue of *Mother Jones* magazine. Under the headline "The Medium Is the Meringue," it described a group of activists who call themselves the Biotic Baking Brigade:

> Targeting the "upper crust"—those they believe to be otherwise "unaccountable" for a variety of corporate crimes—the BBB has tossed pies in the faces (or general direction) of 11 individuals besides [San Francisco Mayor Willie] Brown, including the economist and Nobel laureate Milton Friedman, Monsanto CEO Robert Shapiro, Novartis Corporation CEO Douglas Watson, San Francisco city supervisor Gavin Newsom, Maxxam CEO Charles Hurwitz, and Sierra Club executive director Carl Pope. In addition, the BBB claims affiliation with a "worldwide pastry uprising" that includes the pieing of World Trade Organization director general Renato Ruggiero.

The article profiled the Cherry Pie Three, young activists on trial for "pieing" Mayor Brown. "Nothing they've ever done has brought them as much attention as throwing pies," the article said, noting that the incident made the front page of the *New York Times*. The article quoted one of the young activists as saying, "I think people are going to take up pieing when they read this." She was, in Bob Greenberg's case, absolutely right. Soon, the newest member of the Biotic Baking Brigade would add another name to the BBB's list: Carol Flynn, chair of the Minnesota Senate transportation committee.

Flynn had recently decided to kill the Senate version of Karen Clark's bill by refusing to hear it in her committee. The move, in Greenberg's opinion, warranted a pie. Greenberg had, he explained later, an inside source who told him all about it. In Greenberg's mind, it was perfectly clear: "Carol Flynn had in a backroom deal abused her power and violated the law," he said later. "I got that phone call and an hour and half later I was at the Capitol, pie in hand." He had spent eighteen dollars on it. It was a vegan pie, the kind preferred by the Biotic Baking Brigade.

Later Bob Greenberg would insist that he "gently pressed" rather than threw the pie into Flynn's face, but in any case, the sixty-five-year-old senator fell to the marble floor. To many, it seemed more like an assault than a prank, and other lawmakers were shaken. "It happened so fast, you didn't know what he was taking out of that bag," Senator Ellen Anderson told a newspaper reporter. "It was terrifying."

Linda Brown was appalled when she heard the news. She had still hoped for help from the legislature, even after Senator Flynn had killed the resolution. She was outraged when she recalled that Bob Greenberg had been standing *right there*, listening, as Senator Laidig, the Mendota Dakota's ally from Stillwater, encouraged her and the others to keep trying. It seemed to her that Greenberg had utterly betrayed them. "There was still a little ray of hope, and he killed it. He did it because he wasn't getting enough press," she said bitterly. "So he'd get his name in the paper."

The Mendota Dakota sent flowers and an apology to Flynn, and letters of apology to every member of the legislature, but it was no use. As Karen Clark saw it, there couldn't have been a more effective way of killing the legislation: "Once Bob did that, it was over," she said.

A month later, things did not seem to be going well for Bob Greenberg. He faced charges of intimidating a legislator, disorderly conduct, and fifth-degree assault. The Mendota Dakota had denounced him at a press conference. The coalition of neighborhood groups fighting the highway had kicked him out, and he was no longer welcome at camp. But if he had any second thoughts about the pie, he didn't admit it. "I figured some would distance themselves from me," he said. "It's done exactly what I envisioned it would do: it's drawn a lot of attention." He wore a button with a picture of a flying pie and the words "Sploosh. No remorse."

The pie, Bob Greenberg said, had been "a strategic move to call attention to illegal acts. It was to put some spin on the issue." By most people's lights, all he had accomplished was to create a burst of negative publicity and to end whatever hope the Mendota Dakota had at the legislature—but then, Bob Greenberg didn't think the same way as most people. His world, as anyone who spent much time with him soon learned, ran according to his own Greenbergian logic.

A short, slight man of thirty-one, Bob Greenberg had a sharp-featured, intelligent face and deep-set brown eyes. He dressed casually but conventionally, in jeans and T-shirts. His belief system, not his appearance, set him apart.

He was, he explained, a "primitivist." Primitivists, as Greenberg described them, believe the world needs to be rolled back a millennium or two, to a preindustrial society. Along with industry, agriculture must go too, since it's a form of domination over Grandmother Earth. "A preagricultural society, I believe, is our only salvation," he said.

Ideally, according to the primitivist critique of the world, people should revert to the wild. "Wouldn't it be great to be a naked monkey on a rock, to be really living wild and free?" he asked. He realized that wasn't possible, but he thought the next best thing would be a return to hunting and gathering.

"I've thought about going out and living in the rain forest in the Pacific Northwest, but as a revolutionary I feel I see very clearly the roots of what I see as problems," he explained. Living out in the forest, it would be impossible for him to address these problems. "There's no way I can do it alone." So he was a primitivist in the city.

Greenberg tried to live as lightly as he could on the earth, to buy as little as possible. "You'd be amazed. Take a drive, any evening. Entire Dumpsters are filled with bread," he said. "People throw out so much in this country. Why am I going to put my money into a system I'm doing my utmost to destroy? The greatest strength we have is our refusal to participate, our refusal to give them our money."

Bob Greenberg in happier days at the first protest camp. Photograph courtesy of Mendota Mdewakanton Dakota Community.

Greenberg had a lot on his mind these days. He was convinced the camp would be raided again that weekend, at the end
of April 1999. He had it all figured out. One: the National Guard
was engaging in training missions all around the country. Two:
the powers that be were increasingly interested in, and threatened
by, urban environmental movements. So it would make sense
for the National Guard to use the Highway 55 protest site as a
training exercise. He was almost sure it would happen. He'd
learned a lot about the government and its designs from Internet
sources, which he accessed through an untraceable site. "Some
people think I'm kind of out there," he admitted to a reporter.
"You probably do, too.

"I am a paranoid, you can probably tell that already," he
said. "If you're perceived as a threat to the state, they take action
against you. For me as an organizer, a revolutionary, it's just not in
my best interests for them to see me getting on that [Web] site."

Of the prosecutors in the pie affair, he said, "They're trying to make an example of me. I don't think they understand the
movement." He felt they didn't realize that Earth First! doesn't
have leaders. "Even if they take me out of the picture, they don't
change anything. They just lose one mouth, one strategist, one
activist."

Bob Greenberg knew he was on the fringe of the fringe, the
outback of green anarchy. There were maybe a hundred people
in the world, he estimated, who shared his particular primitivist
views. "I'll be honest with you: not everybody in Earth First! has
a vision that extreme," he said.

But because he was such a ready spokesperson, reporters
often thought Bob Greenberg was an Earth First! leader. "He
was 'leaderized'" by the media, was how Bill Busse, the bearlike
veteran Earth First!er put it. Bill didn't see Greenberg as a leader
at all; he just had the necessary equipment to be a spokesperson.

"He had an apartment and a phone. Bob's house was where the fax machine was." Greenberg was a tireless, highly efficient organizer. He had impeccably ordered computer files and a "blast fax" programmed with newsroom numbers. And when members of the media called him, he called back.

Greenberg's role as spokesperson ended abruptly after the pie incident, but looking back on it, Bob Brown believed it had been a grave mistake ever to have allowed Greenberg to represent the anti–Highway 55 movement. Greenberg had been a public relations disaster from the start. It frustrated Bob Brown no end. One off-the-wall press release he remembered with particular disgust had announced that people in the camp would consent to leave if a series of conditions were met—including an end to sanctions in Iraq. Sanctions in *Iraq!* The guy was a maniac. Brown also suspected Bob Greenberg had used the Mendota Dakota for his own ends. "We just got duped. We didn't know any better," Brown said.

"He's a fundamentally decent guy with some ideas that are a little skewed," said Natalia when she was asked for her opinion of Bob Greenberg. "Bob was always behind the scenes, he did the media work and operated the fax machine. He was never in the front lines, he never got to have fun ... he was hanging out in his apartment while all the rest of us were getting to hang banners and run around. He was the media guy, but that was all that he did."

As for why he threw the pie, "I think he just got sick of being invisible, and just did it because of that," Natalia said. She paused. "I think. I can't speak for him. He made a lot of people mad at the Free State, but he did a lot of really decent things. Like half the camp lived at his house at one time or another, myself included."

Out at the camp, Bear, the taciturn Northern Cheyenne,

added a new button to the ever-changing collection on his army jacket. The button said "Sploosh. No remorse."

"Bob's my friend," Bear said stoutly.

"He tries really hard," Jim Anderson said of Greenberg, "but he doesn't always know when he's hurting things more than he's helping. I mean, nobody knew he was going to do this," Jim said of the pie incident. "Some people just hate him. I don't because I know in his heart he does good work." Jim didn't regret having worked with Bob. "He's really smart. I mean, he does some dumb things like everybody else does, but if it wasn't for his motivation, we might never have found out about Coldwater, or anything else maybe. It's a yin/yang thing. You're going to get good and bad parts out of everybody. I mean, he just had some real piss-poor timing."

As if the pie incident weren't enough, there was another blow— MnDOT's consultants released their report—at the end of April. The report's take-home message was simple: the four oaks weren't old enough to be sacred. The consultants estimated (correctly, it was later determined) that the oaks were only about 130 years old. The report stated that "any trees old enough to have served as Dakota funeral platform supports during the nineteenth century would have to be at least 160 to 170 years old. Historically, it is unlikely that any funeral scaffolds would have been erected in the project area after 1853 (146 years ago), the year when virtually all the inhabitants of traditional Dakota villages near Fort Snelling were forced to relocate to reserved territory along the upper Minnesota River."

The report didn't faze Jim Anderson. Despite the consultants' claim that virtually all the Dakota had left 146 years ago, he knew some Dakota—his ancestors—had remained behind when all the rest went into exile. He thought it was still possible that his own ancestors had used the four oaks for burial scaffolds, even if the trees were only 130 years old. Anderson knew that sometimes Indians did use young trees, supported by dead trees and underbrush, for burial scaffolds.

But even if the four oaks hadn't been used as burial scaffolds, maybe they had served some other function. Maybe, Jim thought, his Dakota ancestors had planted the four oaks as marker trees, as a remembrance of all the Dakota had suffered. (Later, the trees' exact age was determined to be 137 years, meaning they had sprouted in 1862, the year of the Dakota war.) Who could say now for certain what had happened so long ago? All Jim knew was that he had heard the elders say that the trees were sacred, and he believed it.

What really upset Linda Brown was that the consultants seemed to have simply ignored the hours of oral evidence they had listened to so politely. "It's like they're saying that the pope

Watercolor by Seth Eastman of Dakota burial scaffolds on a bluff overlooking the Mississippi River near Fort Snelling, 1846–48. Photograph courtesy of the Minnesota Historical Society.

was lying," she said. "They absolutely did not take into account any of the elders' testimony. It was like, 'Why bother? Why waste these elders' time?'"

The Mendota Dakota had had reason to believe the elders' testimony would be taken into account. Federal guidelines for determining if a site is a Native American traditional cultural property clearly require that oral evidence be considered. The official definition of traditional cultural properties includes sites "associated with the traditional beliefs of a Native American group about its origins, its cultural history, or the nature of the world." Places "where Native American religious practitioners have historically gone, and are known or thought to go today, to perform ceremonial activities in accordance with traditional cultural rules of practice" may also qualify. But when the Mendota Dakota opened the consultants' sixty-eight-page report, they found the elders' testimony summarized in three paragraphs.

The report did acknowledge that the Camp Coldwater spring might qualify as a traditional cultural property. The consultants said it would take more research to determine whether the site met the guidelines. But whether the spring qualified or not, the report said, the highway could be built. The road would not affect the spring, the consultants said, because the nearest on-ramp would be four hundred feet away, and separated from the spring by a hill.

The Mendota Dakota had believed that if they could prove that a patch of green in south Minneapolis was a place where their ancestors held sacred ceremonies and Native people continued to worship, the government would not build a road through it. They believed that the laws protecting Native American religious freedom would stop such a project. They had faith in those laws.

But the Mendota Dakota were not lawyers. If they had been, they would have looked at the legal precedents for some hint of whether the laws actually worked. And they would have found a 1988 Supreme Court case, *Lyng v. Northwest Indian Cemetery Protective Association.* The case concerned a wilderness in the high country of northern California, the Chimney Rock area of the Six Rivers National Forest. The U.S. Forest Service planned to pave a road through the wilderness and open it for logging. "This area, as reported in a study commissioned by the Service, had historically been used by certain American Indians for religious rituals that depend on privacy, silence, and an undisturbed natural setting," the court documents read. Native American organizations and individuals and the State of California sued the Forest Service to stop the road. A lower court judge issued a permanent injunction against it, ruling that it would violate the Indians' constitutional right to freedom of religion.

It is hard to imagine a stronger test case. The land held to be sacred was not a vacant lot in the midst of a noisy city, defended by a tiny group who belonged to no official tribe and had few friends in high places. It was wilderness where for at least two hundred years and probably much longer, according to the court documents, tribes had come for sacred medicine. The lower court had found that intrusions on that particular twenty-five-square-mile area of high country would be "potentially destructive of the very core of Northwest [Indian] religious beliefs and practices." And the tribes were not friendless; they had the State of California on their side.

But the U.S. Supreme Court ruled that the road could go through. The American Indian Religious Freedom Act, "contrary to respondents' contention ... does not create any enforceable legal right that could authorize the District Court's injunction," Justice Sandra Day O'Connor wrote for the five-judge majority. "Even assuming that the Government's actions here will virtually destroy the Indians' ability to practice their religion, the

Constitution simply does not provide a principle that could justify upholding respondents' legal claims."

Justice William Brennan dissented, noting acidly, "The Court believes that Native Americans who request that the Government refrain from destroying their religion effectively seek to extract from the Government *de facto* beneficial ownership of federal property," a line of reasoning Brennan called "astonishing." In effect, he wrote, the court had ruled that it was permissible for the government to use federal land "in a manner that threatens the very existence of a Native American religion." The court had ruled that it was acceptable for the government to destroy the natural places at the very core of Native American religious beliefs; furthermore, the court refused to acknowledge that this was a violation of Native Americans' right to freedom of religion. "This refusal essentially leaves Native Americans with absolutely no constitutional protection against perhaps the gravest threat to their religious practices," Justice Brennan wrote. For Native Americans hoping to save places they held sacred, it was a bleak precedent.

But the Mendota Dakota knew nothing of the Court's decision. They had relied on Mike Haney, who seemed to know so much about federal law and seemed to have all the answers. They had no clue that even if they *had* been able to prove the site was sacred, it would not have done them any good.

Afterward, they called an Indian legal foundation in Colorado, and only then did they learn that the American Indian Religious Freedom Act had no teeth. "That's what they told us," Linda Brown said. "They had never won a case based on the Freedom of Religion Act. Never."

One thing might have saved the land the Mendota Dakota held precious, though: bones. In Minnesota, if road builders or developers find what appear to be long-buried human remains, they must contact the state archaeologist's office. If the remains are determined to be Native American, the Minnesota Indian Affairs Council, a state agency composed of the chairs of Minnesota's eleven federally recognized tribes, decides what to do next. Under Minnesota law, Indian burials cannot be moved without the consent of the Indian Affairs Council. Discovery of Indian burials might have forced the Department of Transportation to put the road somewhere else.

But no bones were found. The field was disturbed ground; its thin skin of weeds and tall grass and wildflowers was only the most recent of many layers. Below it lay chunks of broken concrete and splinters of glass, and beneath that was dark, mucky soil that showed the field had once been a wetland. Ducks had rested and fed there. Reeds and rushes had bent in the wind. Perhaps herons had stalked through on stiltlike legs, spearing frogs. But it was highly unlikely, the consultants pointed out, that Dakota people had buried their dead there. "Nobody buried in swamps," was how consultant John Hotopp explained it to the press.

The wetland was too deeply buried to test for bones. In places, the fill dirt was more than twelve feet thick. Archaeologists with hand shovels dug a few feet from the surface and found no bones, either in the former wetland itself or along its edges.

The Mendota Dakota had failed to make their case. The trees were not old enough. There were no bones. After all the trouble and expense the state had gone to, all the archaeologists they'd hired, "they didn't find a damn thing," as one official put it later.

It seemed there was little left to say. The Mendota Dakota's claims had not stood up to scientific scrutiny. It was as simple as that. Whether the religious beliefs of people in the mainstream would have stood up to the same scientific scrutiny was not an issue.

It's not easy to prove scientifically that a place is sacred. Still, when all the arguments were considered, it was indeed possible that the Mendota Dakota had been wrong. Maybe there *was* no rational basis for all they had believed about the trees and the field and the Coldwater spring. Maybe it all existed in their heads, in a territory outside the rational—the mystical realm we call faith.

At the camp, the four oaks' craggy branches made deep blue shadows on the snow, advancing and retreating like ocean tides each day as the sun rose and set. As the days lengthened, the angle of the shadows shifted. The chickadees' busy, nasal winter calls included new notes: the high, clear whistle that says however cold the days, winter is on its downhill slide toward spring.

By the end of April, the snow had melted for good. In the lawn of one of Riverview Road's vanished houses, someone dug a garden plot and outlined it with stones. It was a beginner's garden of hopeful pale-green pea seedlings, sturdy ranks of radishes sown too thickly, sunflowers planted too early. Seed packets on sticks marked the wobbly rows, each picture bright with promise.

Tarzana, the small-town girl who had lived in the camp all winter, walked by the garden on a mellow spring day, seventy degrees and sunny. She was on her way back from the corner gas station where she had rinsed out her hair, which she had just colored. It came out well, she said happily: minty green and yellow. Before, it had been aquamarine and blond.

She gave a visitor a tour of the camp. Far down the slope of the field, near the wooded riverbank, someone had laid tree limbs in a circle, with streamers of red cloth tied to the branches. Inside the circle, a wood chip path wound in a spiral, a labyrinth with stones piled at its center. Tarzana explained that it was a witches' circle. The witches had held a ceremony March 13, she said. Another circle sat closer to the kitchen, near a newly built bread oven made of brick.

All winter, Tarzana, the stalwart Bear and Caleb, and a few others had kept the ceremonial fire going under the four trees. Now, in the spring air, the scent of wood smoke was stronger. The camp was starting to fill up. People who had left during the winter saying they had problems with camp politics were back. The formerly disaffected activists seemed to have fewer problems, Tarzana observed cynically, now that the weather was warming.

People pitched more tents under the trees and here and there in the field. The plywood and straw bale windbreak around the cooking fire now had a roof of cedar branches. "We did it in camouflage so if they come they can't find us," said Tarzana, laughing.

Witches' labyrinth and cairn, second protest camp. Photograph copyright Keri Pickett.

The camp's impractical dormitory, the Starlodge, now had a third story, pointed like a witch's hat, with jaunty windowed turrets. To reach the third story, people climbed up a ladder leaning on an outside wall. Standing atop the Starlodge in the spring sun—with a wide view of trees on every side and the field sweeping to the river—was like being in a child's tree fort: the same feeling of freedom and limitless possibilities, the same illusion of the grown-up world banished.

Makeshift dormitory known as the Starlodge, second camp.
Photograph by Peter Hoseth.

The big picture window in Bob McFarlin's office at the Department of Transportation had a wide view as well. It overlooked the lawns of the State Capitol mall and the downtown St. Paul skyline.

To many at the camp, McFarlin was the chief minion of MnDOT's malevolent empire and the personification of evil. On the plywood by the entrance to the cooking fire shelter, somebody had mounted a stone gargoyle and misspelled McFarlin's name below it. The real-life McFarlin looked nothing like the stone image at camp, however. The real McFarlin had a thoughtful, open, friendly manner. He had high cheekbones and a patrician look. He wore a polo shirt and khakis. If he personified anything, it was the Republican Party.

McFarlin was occasionally troubled by the public's attitude toward the Department of Transportation. "You have to understand, there's sometimes this public view that people who work for departments of transportation hate trees, hate rivers, and want to pave over the world, and that's simply not true," he said. The department saved all the trees it could, but it was *necessary* to build bridges and roads, he believed. They are the infrastructure society depends upon to function.

McFarlin's father had designed bridges and had been a finalist for Minnesota's transportation commissioner in 1991. McFarlin had transportation in his blood, as he put it. He was an Episcopalian, active in his church. He believed ardently in democracy, in a political system that was fundamentally just and worked well. He believed in what he called "the process."

There was a process for evaluating claims of what was and was not sacred. There had to be. "We live as a community governed by laws to provide some order and balance," he said. "We couldn't function as a community and as a society, we couldn't provide infrastructure services, we couldn't build roads, we couldn't develop networks of communications or all the things that make it a civilization, if every claim of something being of religious value or sacred value or historic value was accepted on its face.

"I have absolutely no qualms about the four trees. I've seen enough evidence. I have read the reports. I have seen the data on the age of the trees. I understand in Jim's heart he believes that those trees served a purpose at a certain time for his ancestors. I think the evidence shows otherwise. Unfortunately for folks like Jim Anderson, they don't like the outcome of the studies. They don't like the facts as they have been discovered, and that's a difference of opinion that I don't think will ever be resolved, frankly."

What really bothered McFarlin were people who went outside the process. People like Bob Greenberg. In meetings with Greenberg and Jim Anderson, McFarlin had told them both from the beginning that they could fight the road by filing lawsuits, not chaining themselves to cement. They could put their claims forward, and those claims would be fairly examined, he said. But Bob Greenberg didn't believe in law or process, as McFarlin saw it.

Once, McFarlin had asked Greenberg whether, if the courts examined the protesters' claims and found them lacking, he would accept those findings. Greenberg, according to the transcript of the tape MnDOT made of the meeting, said the courts had been wrong in the past. "So pursuing your legal claims or a legal path to your claims is really a redundant exercise, because if it doesn't prove your claim out, you won't believe anyone?" asked McFarlin.

"That's correct," said Greenberg.

Greenberg, McFarlin recalled with an edge in his normally pleasant voice, "ardently and eagerly admitted he was an anarchist." McFarlin suspected Greenberg's ambitions went way beyond stopping a road project in Minneapolis: "Bob Greenberg had a wider agenda."

At night, after his commute home to the leafy suburb of Minnetonka, McFarlin would check out the protesters' Web page on the Internet. It was fascinating, really, to flip on the computer and see their view of the world. From links he found there, McFarlin could click into the most extreme fringes of the activist cosmos. "You'd get into the sites that are Bob Greenberg's home," he recalled. It was the terrain of people who didn't believe in the process.

But Bob McFarlin had known from the beginning that the only way opponents could ever stop the highway was through the process. Chaining themselves to cement, listening to extremists, following the Bob Greenbergs of the world would get them nowhere. A judge's ruling or the intervention of the state legislature—avenues that fell within the process—had always been the road opponents' only chance of success, McFarlin believed. He had told Jim Anderson and Bob Greenberg that from the start, when the activists first took over the houses on Riverview Road.

As the snow melted off the wide lawns outside his picture window, McFarlin had just a few more weeks left at MnDOT. He had resigned earlier that spring. By early summer, he would be gone. A new commissioner, Elwyn Tinklenberg, had taken office, appointed by the new governor, Jesse Ventura. Ventura wanted his own people, McFarlin explained.

Asked if the highway fight had taken an emotional toll, McFarlin said that it had. It had dominated his work life ever since August. It had been an "interesting professional experience. . . . I've never managed a situation like this, and quite frankly hope I never do again. It was amazing to me personally the amount of time and emotion expended on this particular project. But I tried, I really tried hard not to get too personally emotive about it."

What emerged most strikingly as McFarlin recalled the highway fight was his own belief in a fair and a just system that gives the same consideration to ragtag bands in cement-block church basements as it does to those who sit in suits around polished conference tables. The process had worked, and worked well, McFarlin insisted. He returned to that point again and again, repeating it like a litany, like an article of faith.

CHAPTER 7

Spring Comes to the Minnehaha
Spiritual Encampment

The days lengthened, and the leaf buds on the four oaks swelled
for the last time. On the weathered branches, new leaves opened
like a pale green mist, followed by tiny flowers dangling from
slender threads. In the protesters' garden, lettuce sprouted in
sparse rows. Someone set out tomato plants and tied them to
branches sunk into the warming earth. A brightly painted plywood sign read "Take what you need and leave the rest."

By late May, the hedge of bridal wreath bushes had burst
into bloom along the driveway to Carol Kratz's vanished house.
Around four o'clock on a warm afternoon, a small fleet of bicycles
lay on the ground or leaned against the young cottonwoods at
the edge of camp. People had come for a meeting, to talk about
the next raid. With winter over, the construction season loomed.

The group sat in a ragged circle, on plastic buckets, coolers,
the camp's motley collection of lawn chairs, or the wood chip–
covered ground. Jim Anderson, Bear, Solstice, Marshall Law
(Solstice's basement lockdown partner during the raid), Natalia,
Dr. Toxic, the six-foot-tall Spiney, and Emily, the shiny-haired
Macalester grad, sat solemnly among a dozen or so others. Bob
Greenberg, who was no longer welcome at the camp, was absent.

The talk turned first to the arrests that would surely accompany the raid. "Is jail solidarity on the agenda?" asked Marshall. It was.

"The code I've heard most is 'No walk, no talk, no sign, no dine,'" said Spiney. "The longer we don't eat, the more scared they're going to be." She had left her dangling test tube earring behind today, but wore a T-shirt that said "If I had a hammer I'd smash the patriarchy."

No one could tell how long they would be in jail, since they didn't know what they would be charged with. Whose land, exactly, were they trespassing on, someone wanted to know.

"Is it government property?" asked Marshall.

"It's no man's land," said Jim.

The main thing when it came to jail was for everybody to stick together against the authorities. "I agree with solidarity," said Solstice. "I think it works. It drives them crazy."

A young girl called Freedom wondered if refusing to answer police questions was really a good idea. "A lot of people make a mockery of me for this, but I really like to treat police as people," she said, a bit uncertainly. A senior at a south Minneapolis high school, Freedom had a mop of curly hair and a faint resemblance to the folksinger Arlo Guthrie. A loyal and active member of the camp community, she got up early in the morning to go on security patrol, went to school during the day, and did her homework at night by the campfire. The others in the circle didn't argue with her. All ideas, however unconventional, were welcome in the mix.

People raised their hands and spoke formally in turn. The meeting dragged on, as meetings do. An ovenbird called from the woods, *Tcher! Tcher! TCHER!* The sun set behind the four oaks. Their leaves shone a translucent green, outlined in gold.

Tarzana slipped into the circle. Her hair was shaved close to her head now and growing back dark brown. She looked like a

Tibetan boy monk. A photographer arrived, prompting Spiney and Marshall to cover their faces, like Zapatistas.

Sky, a twenty-three-year-old Dakota, said that when the raid came there was no point in running around in the woods trying to get arrested, pissing off the cops. Sky had lived at the camp on and off since its founding the summer before. He was a flamboyant dresser, often adding a dramatic flair to marches and demonstrations with his fringed buckskins, painted face, and the feathers he wore in his dark, curly hair. Today he was dressed in full camouflage with a matching camouflage kerchief.

Sky said it was time to get back to basics: prayers, cere-monies, and protection of the Coldwater spring. When the cops came, it was important to have lots of people there praying. "Myself, I'd like to stand outside of a tipi and say to the cops, 'Hey, our elders are inside praying. Can you respect that?'"

"The only reason I'm here is defending the sacred site," said another American Indian man, Ten Bears. "I've been here for a

Strategy meeting at the camp during the spring of 1999. *Front row, left to right*: Marshall Law, Freedom, Spiney, Dr. Toxic, Soil. Photograph copyright Keri Pickett.

while and I like all of you people for what you're doing, but to me the important thing is over there," he said, nodding toward the four trees, "and down the road a ways, the spring."

Bear stood up. "A lot of you people don't know nothing about religion, I can see that right now," he said to the young Earth First!ers. "Some of you don't respect our ways. Half of you people won't even come to our drum ceremony. Some of you people sit on your ass, you won't respect the sacred fire."

Bear said that sometimes he got up at two in the morning and the sacred fire was almost dead, just like it had been when the police put it out during the December raid. "We don't want to see what happened last time, desecrating our sacred fire," he said. He reminded the Earth First!ers how in the raid the police had smashed the Indians' sacred drum. "A lot of us wouldn't sing for months," he added gruffly. When the police came again, Bear planned to be sitting by the fire.

Strategy meeting at the camp, spring 1999. *Front row, left to right*: Sarah, Sky, Bear. Photograph copyright Keri Pickett.

That spring, Emily lived in the camp full time. She had rented out her apartment and cut back to twenty hours a week at the bookstore where she worked. Now she often did surveillance.

To the north of the camp, surveyors in blaze orange vests had been spotted along the old route of Highway 55, parking their Suburbans and unloading their equipment. They set up tripods and aimed laser beams. They drove stakes into the ground. People in camp kept a constant lookout for them. Regularly, Emily took a footpath to the back road behind the Veterans Administration Medical Center complex. There a gigantic cream-colored 1988 Oldsmobile (borrowed from a friend) waited. She fired it up and went on patrol.

The little-used road ran for only a block, between the wild weedy field where the four bur oaks stood to the east and the mowed lawns of the VA Medical Center on the west. It ended

View looking north on Riverview Road, May 1999. Carol Kratz's house and others demolished for the highway sat on the east side of the street. Photograph copyright Keri Pickett.

abruptly at Fifty-fourth Street. She turned left there, away from the river, drove one more block to the old route of Highway 55, then cruised north on the old highway, steering past the corner gas station, a used-car lot, and two cheap motels. She rumbled along the border of Minnehaha Park to where Forty-sixth street crossed the highway. The men in orange vests, if they were around, would be somewhere in this eight-block stretch.

Forty-sixth Street marked the spot on old Highway 55 where MnDOT planned to begin construction that summer. From there, the plans called for the construction work to move south, toward the protest camp.

When she got to Forty-sixth Street, Emily turned the Oldsmobile around by the strip mall on the east side of the road. An odd landmark stood in the mall parking lot: a twenty-foot-tall kitsch imitation of a Northwest Indian totem pole. A Thunderbird with wings outstretched perched atop a neon sign saying PARKWAY PLAZA, supported by a pole of fake-carved, stacked faces.

If you looked north from the strip mall, you could see the finished, new-and-improved stretch of Highway 55 to downtown Minneapolis. At only four lanes, the highway wasn't nearly as wide as MnDOT had once planned. Beside the finished highway, the land where hundreds of houses had been razed for an eight-lane freeway stood empty, a barren no-man's-land edged with a sound wall and rows of spindly trees. But at least that stretch was completed. Now MnDOT was poised to move south and finish the job. The activists watched and waited.

Emily cruised back toward the camp. No surveyors out just now. She knew where to look for them. She knew exactly where the road was planned, where it would obliterate the shade trees and lawns on the edge of Minnehaha Park. She could picture where it would swing away from Highway 55's existing route— carefully preserving the seedy motels, used-car lot, and gas station

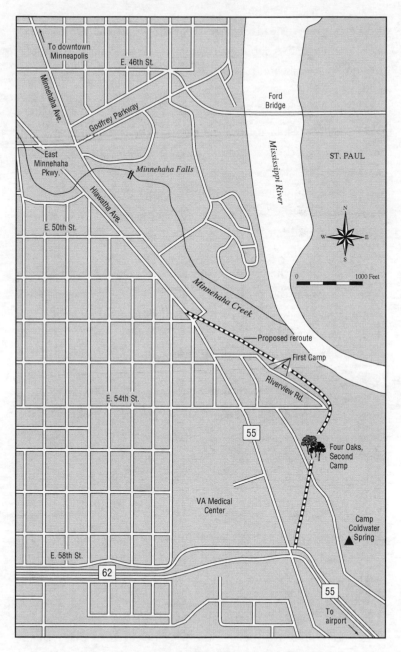

Map of camp, oaks, and eight-block stretch of old Highway 55.

and destroying instead the green swath along the river. When other people had trouble picturing it, Emily offered a simple, through-the-windshield guideline: "Wherever there are trees, that's where the road goes."

Emily still believed it could be stopped. "Maybe there won't be clean resolution and I'll have to find my own resolution, but right now I think it can be done," she said. She was twenty-three years old. Stopping the road was immensely important to her. "There's only one moment in time when I can be involved in an issue like this," she said. At the bookstore, one of her co-workers had told her, "This might be the most important thing you do in your life."

She parked the Oldsmobile on the road and walked back to camp. A sign by the path said "Mendota Mdewakanton Dakota Community." The Earth First!ers had erected their own sign: "Footpath. To where, you ask? To the trees."

It was hard to say exactly how many people lived in the camp that spring—maybe twenty; at most thirty. Many of them were from a "traveler" culture, Emily explained, an underground movement of people who spent their lives without jobs or apartments. They came and went. Trying to put a number on them was, she said patiently, like "taking a picture of a collage as somebody is making it." They were hanging on; that was what was important.

"On August 10," Emily declared in a speech in Minnehaha Park a few days later, "the encampment we thought would last maybe a day or two will celebrate its one-year anniversary." There would be new lawsuits, she explained. People were doing "righteous work."

"We're operating on a continuing success basis," she said in her studious way. She was more clean-cut than many of the others at camp. She went by her real name, and she seemed more eager to please than someone like Spiney, who simply brushed off questions about her background. "I could have been a stockbroker yesterday and if I'm here today, I'm part of the movement," Spiney said when she was asked.

Emily finished her speech and sat down on the grass. Spiney stood up next, dressed in a black mosquito costume. Long cardboard wings hung from her shoulders and a mask with a proboscis and bulging papier-mâché eyes covered her face. The surveyors were out in the neighborhood now, she said, but they hadn't been able to do their work. (In the days before, people had played a lot of Frisbee in the surveyors' line of sight. At night, stakes had been disappearing, somehow.)

"We're out every day stopping them," Spiney told dozens of people listening in the grassy shade. She invited them to come on out whenever they saw the yellow tripods and orange Suburbans. "The mosquitoes are small but there's a lot of 'em," she said. "All the chemical sprays in the world can't keep them away."

"Tell it, tell it!" called a voice from the crowd.

CHAPTER 8

Tree by Tree

When Nettle first heard about the Highway 55 fight, she was herding sheep at Big Mountain, a remote stretch of high desert in northeastern Arizona. Nettle was a "land supporter," one of a group of activists who chop firewood, haul water, and help Navajo elders maintain their traditional shepherding way of life. Nettle and the other activists were there to support the elders as they resisted the efforts of the federal government and the Hopi Tribal Council to remove the Navajo from disputed lands.

Nettle had met Marshall Law and Sky, the flamboyant twenty-three-year-old Dakota, when they traveled to Big Mountain, bringing news of the fight in Minneapolis and saying they might need help when construction started. "We told them to let us know if there came a time when there was a red alert, when a lot of people would be needed," Nettle remembered. "We said we would come right away if we got that message."

One July morning, Nettle and her partner, Squash, woke up and had a strange feeling: something told them it was time to go to Minneapolis, as Nettle explained it. "So we hit the road," she recalled. They caught a ride most of the way with some people who had come to the Sundance at Big Mountain, then hitchhiked the rest. Halfway to Minnesota, they checked somebody's voice mail and learned the red alert was on.

Standing in the parking lot by the Parkway Plaza totem pole, Solstice could see squad cars converging. The heat radiated from the asphalt under his feet; the sun glinted off the glass and steel strip mall. The giant Burger King sign glared.

But across the highway, where the squad cars had pulled onto the grass, it was still cool and green. Elms two stories tall, silver maples, and small cedars shaded lawns sloping down to Minnehaha Creek. The area, a backwater where Minnehaha Creek took a sharp curve, was known fondly in the neighborhood as the Longfellow Lagoon. Although it was separated from the rest of Minnehaha Park by the highway, Longfellow Lagoon was officially part of the park.

The little patch of park around the lagoon sat about half a mile north of the camp and the four bur oaks. It was the first place slated to be cleared for the widening and rerouting of Highway 55. Now workers had enclosed the area, which was the size of several football fields, with orange plastic fencing. Officers stationed themselves every few feet along the orange line. Inside the fencing, on the shady lawn, a backhoe and a front-end loader were poised and ready, flanked by more squad cars and a paramedic van. Solstice began calling supporters on his cell phone.

MnDOT commissioner Elwyn Tinklenberg had announced the start of construction that morning, July 26, at a Capitol press conference. "The people who live along the corridor, who have been subjected to the uncertainty and disruption this project have entailed, deserve closure," Tinklenberg said. State troopers, Minneapolis police, and Hennepin County sheriff's deputies were on the scene, "to protect the safety of our workers and our construction personnel and the general public," he announced. "This is beginning as we speak."

Outside the orange fence, the crowd grew quickly. Jim Anderson, Bob and Linda Brown, Tumbleweed, Tarzana, and dozens of others came as soon as they heard. They waved signs: "America, where's your sacred land?" and "Bury my heart where MnDOT won't find it." They chanted "Stop the reroute!," taunted the police, and milled against the fencing. Suddenly a teenage girl slipped through and sprinted toward the trees. A cry rose from the crowd, an eerie, ululating wail: *luh luh luh luh luh!* Officers tackled the girl and brought her to the ground. Another person broke through the barrier and ran, then another. *Luh luh luh luh luh!* wailed the crowd.

Soon the first elms were falling, their green tops arching through the air and crashing to the ground. Each time a tree toppled, there was a lull in the yelling, a shocked pause while chain saws whined and a State Patrol helicopter thumped overhead. A bulldozer set its blade against the trunk of an elm tree four feet in diameter. A man burst from the crowd and leaped toward the machine. A clump of officers brought him down. "Tumbleweed! Tumbleweed!" called the crowd. The elm heaved over.

In the blazing heat, the shade was soon gone from the lagoon. By afternoon the crowd of protesters huddled under a scraggly stand of scrub trees outside the orange fence. By the day's end, the lawns were heaped with sawn trunks and branches. A state trooper in his brown uniform and Smokey the Bear hat stood next to a fallen elm; the width of the cut trunk reached to his waist. The park had the harsh, bright look of a shopping mall parking lot, where the sun glares back up from the pavement.

In the air-conditioned chill of the Subway sandwich shop across the highway, diners discussed the spectacle in the park. A man in a baseball cap had seen some guy swimming Minnehaha

Creek trying to get to the trees before the cops got him. Unbelievable what some of these people would do, he said. Somebody else said there was no need to cut down all those trees, but anyone could see it was impossible to stop: "If the government wants to, the government's going to do it."

Two days later, the cut trees had vanished. They had been ground into mounds of wood chips twenty feet high, which now dotted the lawn by the creek. A dozen or so people stood nearby, holding signs saying "Honk for the trees." An occasional motorist yelled "Get a job!" but on that July morning, there were more honks than jibes. Not just from beat-up, bumper-stickered, Nader-for-president cars, but also from shiny new ones. Suburban commuters leaned on their horns. The gaggle of protesters raised their fists.

In the days after the first cutting, the list of arrests grew. Solstice, Tumbleweed, and seven others were arrested the first day. A day later, Jim Anderson, Bear, Tarzana, Marshall Law, Sharon Day, and more than a dozen others linked arms and stepped out into rush-hour traffic on Highway 55. Within a week, the number of arrests topped fifty.

Crews with chain saws began felling trees along the soon-to-be-widened stretch of Highway 55 between the totem pole and camp. A yellow bulldozer-sized timber harvester called a Hydro-Ax began grazing along the highway's path, leaving clusters of stumps behind.

The activists' fight against the highway ratcheted up a level. "That's when we decided, instead of just sitting [in camp] waiting for them to come to us, let's try to make it a little bit tougher for them," Dr. Toxic recalled.

The Earth First!ers posted a sentry car with a CB radio near a huge cottonwood overhanging a footbridge spanning Minnehaha Creek. They bicycled up and down the highway doing security. Fleets of Earth First!ers would circle the big equipment, then somebody would slip past the cops, hop onto a machine, and fasten himself or herself to it with a U-shaped bicycle lock. There were dozens of such skirmishes.

The big shade trees along the highway were soon festooned with hammocks, sleeping bags, dangling plastic water jugs, and nets of food. A banner in the cottonwood by the creek asked in foot-tall, hand-painted letters, "Is economic progress killing the planet?" From up in the trees, people with CB radios watched the construction crews, and the crews watched back.

Judy Melander, who had replaced Bob McFarlin as MnDOT's main spokesperson, sometimes visited the site to talk to reporters. One day toward the end of July she stood near a cluster of shade trees along the fringes of Minnehaha Park. She stationed herself at a slight distance from a dreadlocked guy who had chained his arms around a tree trunk. Nearby, a fifteen-year-old girl looked down from another tree's branches.

Melander had been a spokesperson for various government agencies for years, but she'd never had to deal with anything as unpredictable as venturing out to the scene of the Highway 55 protests. Once she had to cut short a television interview when she glimpsed someone maneuvering into camera range behind her holding a sign saying "MnDOT Daughter of Satan." She found it hard to know how to dress for such occasions, but today she wore brightly colored culottes, tennis shoes, and a sun visor. Already, she was wilting in the heat. She pointed out that the protesters were costing the taxpayers big money. Police protection for the construction site was running $50,000 a day, according to MnDOT. Adding in the law enforcement agencies' equipment costs, the tab was running at $60.10 an hour per officer.

At three-thirty the next morning, in a warm summer rain, Marshall Law trudged along the road between the VA Medical

Center and the camp. His T-shirt and shorts were drenched. In the orange glare of the hospital's lights, his bearded face looked gaunt. Somebody had seen squad cars on the move and what looked like paddy wagons parked by the shade trees along the highway and had alerted the whole camp. It turned out to be a false alarm, Marshall said, but it had been eerie as he stood watching the dark shapes in the rain.

"It's been a hard couple of days," Marshall said wearily. "We've taken a lot of injuries." Madeline Gardner, a slender seventeen-year-old, had gone to the hospital with a dislocated hip after the police tackled her. She was on crutches now. Emily, arrested a day later, had her head banged on a railroad tie.

(Emily's college friend Sarah had been shocked when she visited Emily in the hospital. "When I saw her, I was really taken aback. Her face was so swollen and her lip was really fat," Sarah said wonderingly. "She had two or three stitches on her forehead and bruises on her scalp underneath her hair.")

Things were getting tougher and tougher, Marshall said. People who had been arrested once now risked another arrest just for coming near the construction site. Solstice had been nabbed in the parking lot near the totem pole. And it was a huge task to try to guard so many trees, spread out over such a big area. "It's a hard one to call. They just keep us stretched out," he said. His handheld radio crackled, and soon Bear's old, white, bumper sticker–encrusted Buick appeared out of the darkness, cruising slowly down the gravel road along the edge of camp. The two conferred, Marshall left to get some sleep, and Bear climbed back into the Buick and rumbled into the night.

The rain let up, and crickets whirred in the tall grass under a full moon. Under the four oaks, the ceremonial fire, now sheltered in an iron kettle, burned steadily. Beyond the oaks, torches and a kerosene lantern lit the way to the kitchen. With its firelight and soft shadows, the camp seemed far away from the city's harsh sodium lights and electrified, bruise-colored sky.

A middle-aged woman named Pam sat next to the cooking fire in the straw-bale kitchen shelter. "Shoot, you missed the hot dogs," she said to a visitor. "We had hot dogs and Pop-Tarts at two." The paddy wagon scare had roused the whole camp. "We got all excited and went on full alert again." She sighed.

"Bear coming off the hill," said a voice on her two-way radio.

"Copy that, Bear," she replied.

Pam described herself as Ottawa and Cree, an ordinary person, "not your flaming radical." Three months before, she was a medical claims administrator in Illinois. "I had a good job, a nice little house, the whole nine yards," she said. She decided to come to camp after hearing Jim Anderson speak at an American Indian Movement meeting and seeing his video of the December raid: "I decided it was something I had to do. I'm tired of Indians being routed off their sacred land." She told her visitor that the people at camp were the best she'd ever met—"People that live in thirty-degree-below weather and rain, hundred-degree weather and bugs, just for ideals, I think that's pretty wonderful." She was baffled that the neighbors didn't stand with the camp against MnDOT. "They're going to turn a beautiful green neighborhood into a slum. This is going to be like the South Side of Chicago when they ran the Dan Ryan [Expressway] through it."

Ten Bears, one of the Indian men who lived in the camp, stepped into the kitchen and checked the chalkboard to see who had the next watch: Hillside, Broccoli, and Earthworm. Ten Bears left to wake Earthworm, who was sleeping in the bus. Pam headed off to bed, wondering where the rest of the radios were. "We've probably got ten radios here and nobody knows where they are," she grumbled.

It was quiet for a moment. The daily commute, a world

away on the highway, wouldn't begin for a few hours. For a brief time, no planes roared overhead. Firelight flickered against the makeshift shelves of canned goods, rice bags, and plastic buckets that lined the walls of the kitchen shelter. On the chalkboard, someone had written "Wes! Wake Freedom top of Starlodge!"

Another notice read "Gimme ma boots! Brown leather tall boots near kitchen. Please give to Elaine. Xoxo Squash."

"Don't put unwrapped meat in da cooler, you fools!" said another. To that, someone had added a less mundane piece of advice: "Never give up."

At six the next evening, it was still and ninety-four degrees. The Mississippi lay as flat and glassy as a lake. In the field by the four trees, goldenrod was just coming into bloom. A dry, hot sagebrush smell hung in the air.

Nettle from Big Mountain passed by the kitchen, wearing a nylon rock-climbing harness, carabiners dangling from it like oversized paper clips. (Carabiners are a climber's tool, used to snap the harness to a safety rope running up a rock face.)

In the tiny oak forest between the camp and Riverview Road, ropes were strung from tree to tree, connecting a series of hammocks and platforms made out of pallets. If the cops came after you in one tree, you could slide to another without touching the ground, a barefoot teenager named Shane explained. "It's not really the point to piss off cops, but it's an added plus," he said, grinning.

Shane, Nettle, and Squash had arrived from Big Mountain a few days earlier. Shane had made the trip in just two days; one "slick ride" in the back of a pickup truck had taken him through all of Nebraska. Now, ignoring the heat and bugs, Shane and Squash strolled cheerfully down the path known in camp as the Noble Oaks Trail.

Squash's appearance was startling, even for an Earth First!er. His nose was pierced through the septum like a bull's; sometimes he wore a silver horseshoe-shaped ring in it, other times a bone pointed at each end. With his curly brown beard and long dreads, he looked like a faun, or like Puck in *A Midsummer Night's Dream*. Not everybody at camp looked like that, Shane pointed out. "There's people here who are, like, clean-cut truckers," he said.

"And then there's us," said Squash, laughing.

Later, as Shane and Squash and half a dozen others sat around the campfire, they tried to explain their wandering life to a visitor. They had participated in "actions," not just helping out the Navajo elders at Big Mountain, but in other places, too, they said. "There are struggles going on all over the country. We travel and teach each other," said Squash.

They joked about their subculture, explaining its peculiarities as though they were Samoans instructing Margaret Mead. Carhartt work clothes were the latest thing in Earth First! fashion, they deadpanned. The all-purpose tool known as a Leatherman was very big—quite useful for various direct action activities. "MacGyver in a box," somebody called it. (MacGyver, the ingenious hero of a 1980s television show, used a Swiss army knife, duct tape, shoelaces, and the like to escape from danger.)

"A lot of subcultural movements have contributed to the amalgamation of weirdos you see here," said Wes, mock-seriously. Wes was twenty-two, with brown shoulder-length hair and a beard that made him look older. He wore heavy dark-rimmed glasses that seemed like a leftover from some former life. Back in the December raid ("three or four hundred years ago," he joked), he had been the guy dressed as Santa Claus.

Wes was a habitual wanderer, often hopping freight trains on various missions about which he was deliberately vague. "Fomenting dissent," he joked if you asked him. He passed through the Free State every so often. Last winter, in twenty-below cold, he and Spiney had rigged up the new camp's first tree sit. He helped build the Starlodge but didn't stay long afterward.

Now he had just come from Alaska, where he'd been working on a campaign to stop a road from being paved across the Copper River delta. You had to go where you were needed and do whatever was right for the time and place, he explained. You had to be a "liquid warrior."

The group around the campfire was not a bunch of "cheesy pot-smoking hippies," Wes said. It was a manifestation of a "broader black and green ecoanarchist consciousness," he said, half joking, half serious. He answered questions courteously but in a mocking tone, as if parodying a college professor. But his message was clear: the amalgamated weirdos were part of something bigger than themselves.

It was a *movement*, not an organization, as anyone who tried to explain it would tell you. It had no official membership, no paid staff, no hierarchy, and no leaders. It resisted labels. It wasn't even that crazy about being called Earth First! because it was more than that. But like any other movement, it had a history. And the story began with the Earth First! founding.

It was 1980. Ronald Reagan, whose attitude toward wilderness was perhaps best summarized by a legendary remark—"If you've seen one redwood, you've seen them all"—occupied the White House. James Watt, who favored opening the nation's entire coastline to oil drilling, was secretary of the interior. And Dave Foreman, who would soon be one of the founders of Earth First!, was fed up. A beer-drinking, cowboy-booted westerner who didn't mix well with the environmental establishment in Washington, D.C., he decided to quit his lobbying job with the Wilderness Society. In his opinion, mainstream environmental groups were losing the fight against Reagan and his big-business friends. Environmental groups were caving under the pressure, Foreman thought, letting too many wild places fall to mining and development. Millions of acres of forest that could have been forever preserved as wilderness were open to logging because the mainstream groups had cut too weak a deal with the U.S. Forest Service. It was time to do something different.

Around a campfire in the stark beauty of the Pinacate Desert in northern Mexico, Foreman and a few friends hatched the idea of an environmental group that would make no compromises. Its logo would be a raised, clenched fist. It would take its lead from the fictional ecosaboteurs in Edward Abbey's novel *The Monkey Wrench Gang*, who sawed down billboards, toppled power line towers, incapacitated bulldozers, and dreamed of blowing up the Glen Canyon Dam. Foreman and his friends chose the same dam for the group's national debut.

To the Earth First! founders, the Glen Canyon Dam was the ultimate example of the devastation caused by environmental compromise. As part of a deal to save the Grand Canyon and other parts of the Colorado River, the Sierra Club had reluctantly agreed to a giant dam on the Arizona-Utah border. Completed in 1963, the dam backed up the Colorado for 186 miles, drowning the spectacular Glen Canyon forever. The impounded river formed Lake Powell, named after Major John Wesley Powell, the first American to systematically explore the canyon. "Where he and his brave men once lined the rapids and glided through silent canyons two thousand feet deep the motorboats now smoke and whine, scumming the water with cigarette butts, beer cans and oil, dragging the water skiers on their endless rounds, clockwise," Edward Abbey wrote.

In March 1981, Foreman and four others slipped past police to unfurl a three-hundred-foot plastic banner—a symbolic crack—down the dam's vast cement face. From a nearby bridge, sixty or seventy supporters cheered, and Edward Abbey himself yelled "Earth First!" and "Free the Colorado!"

"It is not enough to protect our few remaining bits of wilderness," Foreman wrote in a 1981 article in the *Progressive*. "The only hope for Earth (and humanity for that matter) is to withdraw huge areas as inviolate natural sanctuaries from the depredations of modern industry and technology.... Move out the people and cars. Reclaim the roads and plowed land. It is not enough any longer to say no more dams on our wild rivers. We must begin tearing down some dams already built—beginning with Glen Canyon, Hetch Hetchy, Tellico, and New Melones— and freeing shackled rivers."

The goal of Earth First! was to create an unabashed radical wing of the environmental movement. Earth First! would draw a new line in the sand, in a far different place than the Sierra Club had. In part, it was a conscious effort to recast mainstream environmental groups—labeled extremists by the Reaganites—as middle-of the-roaders. Earth First! would make the Sierra Club and its allies look like "raging moderates," Foreman wrote.

Although Earth First! itself would be "ostensibly law-abiding," Foreman and rest of the group's founders agreed, it would try to "inspire others to carry out activities straight from the pages of *The Monkey Wrench Gang*." Foreman's 1985 book *Ecodefense: A Field Guide to Monkeywrenching*, became the movement's how-to manual.

From the beginning, Foreman and the others who conceived Earth First! envisioned a do-it-yourself movement. Anyone who wanted to could claim the Earth First! label and take whatever action they deemed appropriate. "I think we had the self-restraint to let Earth First! develop itself," Foreman told journalist Susan Zakin. The idea was "to let go and trust other people, to say, 'OK, start a local Earth First! movement. Go for it. Here are some ideas but use your creativity,'" Foreman said.

"The group's anarchistic makeup . . . helped it spread around the country like a weed," wrote Zakin. Soon, people in Texas, California, and New England were calling themselves Earth First!ers.

"It was only a couple of years ago when there were four or five of us in a car and someone came up with the thought that if we got squashed by a semi truck there would be no Earth First! movement," Foreman told Zakin in May 1986. "We're at the point now where it would take a much larger vehicle," he joked.

The precise number of Earth First!ers was impossible to count. In the second half of the 1980s, the *Earth First! Journal* had about five thousand subscribers, according to Zakin, who estimates that at its high point the group attracted five thousand to ten thousand followers.

Some in the movement heeded the Earth First! call to ecosabotage. In Wyoming, monkeywrenchers destroyed seismographic equipment and pulled survey stakes to stop oil exploration in National Forest land south of Yellowstone National Park. In the Pacific Northwest, they set fire to timber company bulldozers or slashed the machines' hydraulic hoses and wiring and poured abrasives into the engines. By 1984, they had begun driving metal stakes into trees.

The practice, known as "tree spiking," was controversial even within Earth First! because of the risk that spikes pose to loggers and sawmill workers. Supporters of tree spiking insisted the goal was not to hurt workers but to drive up the cost of logging by forcing timber companies to remove the spikes before trees could be cut. For that reason, the recommended Earth First! method was to post warning notices where trees had been spiked. But although reported injuries from tree spiking were few, in one confirmed case, a California millworker was seriously hurt when his saw struck an eleven-inch spike.

Precise figures were never compiled, but according to some estimates, by the late 1980s tree spiking was costing timber companies twenty million dollars a year. (In 1987 the U.S. Forest Service commissioned a survey of its personnel nationwide to estimate the extent of ecosabotage, but it declined to release the results to the public.)

Earth First!ers blockaded timber roads and in 1985 pioneered a new tactic: using rock-climbing equipment to scale giant trees and set up residence. In the first tree sits, two Earth First!ers tried to stop the logging of cathedral-like stands of thousand-year-old Douglas firs in Oregon's Willamette National Forest. A 1989 *Los Angeles Times* headline read "Radical Environmentalists in Trees Disrupt Logging across West."

The effect of such actions was hard to gauge, but without question, the movement did save some wilderness. Earth First!ers who pulled the survey stakes in Wyoming delayed Getty Oil's road long enough for an administrative appeal to halt it for good. (Congress later preserved the area as wilderness.) In southwest Oregon's Kalimopsis roadless area, one of the west's most ancient and diverse forests, an Earth First! blockade in 1983 held off construction of a road for months—long enough for a judge to rule that the proposed road was illegal.

Even less successful Earth First! protests managed to draw attention to the clear-cutting of the nation's ancient forests. The movement's supporters argue that by injecting passion into what had been an arcane issue, radical environmentalists prodded mainstream groups into action. National environmental organizations that for years had been reluctant to alienate timber-state representatives in Congress, the argument goes, finally pressured the U.S. Fish and Wildlife Service to hold hearings on listing the northern spotted owl as an endangered species. After the spotted owl was declared endangered in 1990, a federal district judge ruled huge portions of its old-growth forest habitat off-limits to logging.

But as the 1980s faded into the 1990s, Earth First! seemed to run out of steam. One drawback to its leaderless, do-it-yourself model was that anyone could do something stupid—such as target a responsible family logger instead of a corporate bad guy. Then they could hang up an Earth First! banner, and Earth First! would get the blame. And even when Earth First! activists were on track, their tactics no longer seemed to be working.

In northern California, the Maxxam Corporation was busily felling giant redwoods and exporting the logs to Japanese mills. Earth First! hung a banner over the Golden Gate Bridge and held massive demonstrations during the "Redwood Summer" of 1990—but failed to save a single tree.

And the personal risks of the Earth First! brand of activism were becoming clearer. In the spring before Redwood Summer, a pipe bomb exploded in a car driven by two Earth First! organizers. One of them, Judi Bari, suffered a broken pelvis and permanent nerve injuries. A year earlier, in 1989, Dave Foreman and three others were charged with conspiracy to sabotage nuclear facilities in Arizona, Colorado, and California. In the desert outside Prescott, Arizona, FBI agents had surprised two men and one woman cutting the legs off a power line tower with a blowtorch. Foreman, although he was not present, was charged as a co-conspirator. All four plea bargained. Foreman's codefendants received sentences ranging from six months to six years. Foreman was given a suspended sentence and the chance to plead guilty to a misdemeanor, but his days as a charismatic spokesperson for Earth First! were over.

Without Foreman, Earth First! no longer attracted the

attention it once had. *Outside* magazine's 1991 report card grading fourteen leading environmental organizations gave Earth First! an incomplete. "A grand total of 75 believers attended this year's annual rendezvous in Vermont, which isn't exactly surprising," *Outside*'s editors commented cattily, noting that Foreman had quit the group, media interest was waning, and old-growth forests were a "well trod issue." By the end of the 1990s, the number of subscribers to the *Earth First! Journal* had dropped by more than half, from five thousand to around two thousand.

Still, the movement hung on.

Old-growth forests might well have been old hat to the editors of *Outside* magazine, but timber companies continued to fell ancient trees, and activists using Earth First! tactics still fought them, although the new generation of activists didn't always use the Earth First! name. In Oregon, people who called themselves Cascadia Forest Defenders lived in treetop platforms in the same national forest where Earth First!ers had perched in 1984. Now they called their tree sits ewok villages, after the tree-top homes of the koala-like creatures in a *Star Wars* movie.

By the 1990s the movement had spread abroad, to antiroads actions in England and other countries. In December 1997, Julia Butterfly Hill climbed into the branches of a towering California redwood and began a two-year vigil that attracted international attention.

People still gathered at Earth First! "Rendezvous" (named after the Old West get-togethers of Indians and mountain men) in remote places; they drove battered cars or hitchhiked from one base camp or "free state" to the next. Earth First! persisted as an obscure subculture, like the lifelong fans who traveled with the Grateful Dead. Earth First! "has an amazingly large bark for the number of people who consider themselves Earth First!ers," said James Bell, a veteran of the movement who later went to work for an environmental communications firm in Chicago.

According to Bell, the Earth First! subculture is hard to trace, even on the Internet. "You've got to remember that in this crowd there's a Luddite philosophy," he explained at an Earth First! Rendezvous in southern Minnesota in early 1999. "The *Earth First! Journal* had no Web site till a year ago." Earth First!ers' e-mail (and actual) addresses, when they existed, had a way of turning into dead ends.

Earth First!ers often learned of the group's various encampments by word of mouth. Encampments might be in the roadless areas in northern Idaho, or along northern California's "lost coast," which can be reached only by dirt roads. On the lost coast, marijuana growers would toss down a wad of cash to pay for carabiners or climbing harnesses or protesters' bail. "Marijuana growers want it to stay the lost coast," Bell explained.

Bell lived from 1992 to 1996 with other Earth First!ers in support of Western Shoshone Indians in Nevada trying to stave off gold mining and the federal government's efforts to confiscate Indian livestock. At night around the campfire, he listened to Shoshone elders tell creation stories, or to fellow Earth First!ers joke about botched actions: train-hopping fiascos, or the time somebody read the map backward.

Two decades after its founding, Earth First! was a scraggly tribe with its own history, heroes, folklore, and music. Although the next generation of Earth First!ers worried about new threats— genetic engineering and the global power of multinational corporations—they still believed in many of Dave Foreman's old principles. Around a remote campfire, they could watch the stars wheeling across the night sky and imagine a quieter, greener world with more wilderness and not nearly as many cars.

Around the fire in Minneapolis, the talk turned to what it's like to sit in a tree day after day. "A tree has veins, it has pulse, it moves with the wind," Squash explained. The activists talked of watching woodpeckers and hummingbirds among the branches. Shane had seen a squirrel in its nest. Wes had observed ants marching over him as though he were part of the tree.

Wes was barefoot. He had been missing his shoes for several days. Now, as the talk around the fire wandered, he noticed that Squash was wearing the missing shoes. Wes remarked upon it without rancor, as if it were simply a mildly interesting fact. Squash had grabbed them once when he had to climb a tree fast, it turned out. It didn't make much difference anyway. They both belonged to the same tribe.

In late August, sunflowers nodded over the garden on Riverview Road. Purple eggplants hung like lanterns. A small green pumpkin showed veins of orange. The oak forest now was tinged with yellow, the first few leaves starting to fall. Along the edge of the field, the sumacs flamed red.

Near the four trees, Jim Anderson stood in the warm wind, before a crowd of perhaps a hundred people who had walked to the field from north Minneapolis. The march had been billed as a display of solidarity between two neighborhoods: Minneapolis's poor north side and the working-class south side of town, where the camp stood. The hike had grown out of efforts by Marshall Law, Solstice, and others trying to forge a coalition between the two parts of the city. That morning, some 250 people had started on the walk. Jim thanked the hikers for their support, his voice amplified by a bad public-address system.

"Hau mitakuyeowasin," he said, the Dakota words meaning "Hello, all my relatives." His voice sounded tinny and far away. "There's too many people, especially the people with the money, they're blindsided by that money, they can't see any farther than their own stinkin' lives, you know. What they can get out of this life and the money they can get for themselves. Well, we look at seven generations ahead, we have to look at our kids, because our time here is so short. We'll be gone, and what are our kids going to do if we take their air from them by killing their trees, if we take their water from them by destroying that water, they don't have a future."

Jim spoke without notes, as he always did, one sentence running into the next. He wore plaid shorts, tennis shoes, and a faded T-shirt with "MMDC Security" on the back. "This Earth, she'll come back, I don't know what will be here, but it won't be man, because we're going to destroy ourselves. There's a lot of things that have gone extinct already, and that's what we're on the road to," he said. "If you don't believe it, look around. Look at the catastrophes we're doing by this global warming, by these floods out east, by these major, massive storms, by the earthquakes that are killing people left and right. We gotta do something, and this is the beginning of it."

Native American families, black-clad Earth First!ers, older die-hard activists, and conventionally dressed white people from the neighborhood listened quietly, sitting on straw bales and blankets and the camp's usual eclectic seating. "You know, they say the year 2000 is a time that the world's gonna end. Well I don't believe that," Jim said. "I believe that we're gonna finally pull our heads out of our butts and realize that when we cut ourselves, we're all red inside. We're all people and our children are the ones that are important. That's what this fight's all about. And if you remember that, you can fight on."

Another speech, then everyone moved forward for the pipe ceremony. "We need a full circle, we got more coming in, so just move it down, kinda scuttle on over," Jim said. This year he had been a Sundancer for the first time, he told the audience. He and some others had held a sweat lodge ceremony on Pike Island the night before the march and had prayed for things to go well. "I know there is power in prayer. It happens a lot. Things happen out there."

When the pipe ceremony ended, about half the crowd

formed a line standing shoulder to shoulder. The rest formed a second line, and the two groups passed by one another, shaking hands and chatting and hugging like people in a receiving line after a wedding. Behind them, the camp looked like a movie set. Two horses grazed among the tipis under the trees.

The horses were on loan to Sky, the flamboyant young Dakota man. That fall, Sky and his brother had ridden them to the camp from the border of the Pine Ridge Reservation in South Dakota, seven hundred miles in thirty-eight days. Sky had put up his car as security for the horses. Now, dressed in fringed leather, his face painted, Sky rode one of the horses up a dry grassy ridge set against a background of oaks tinged with autumn. He paused at the crest and raised a staff dangling with hawk feathers. "Like *Dances with Wolves*," remarked someone in the crowd. Sky galloped down the ridge. The onlookers cheered.

The idea for his long ride had come to Sky during a vision quest under the oaks, in the cold October of 1998, two months after the camp began. Sky had fasted under the oaks for four days before the vision came to him. Asked about it much later, he struggled to explain what a vision quest is. "It's like we cry for a dream," he said. He wrote out the word: *Hemblecha*. In Sky's dream, the land was saved. Where the highway would go, there was instead a bridge, its span skimming the crowns of living trees. But looking closer, Sky saw they weren't really trees but cement pillars made to look like trees, and people were painting them—"a good form of graffiti," he said. Around the pillars, he saw waving prairie grass and horses running and buffalo grazing. He still believed in his dream. "That's possible," he said.

On a late August morning in camp, the city could seem far away, despite the planes roaring overhead and the perpetual white noise of cars on Hiawatha Avenue. People would wake to the sight of tall grass and the trees in steadfast ranks around the edges of the field. "Are you heading out into the real world today?" one person would ask another.

But now when they walked to the field's northern border and stepped onto the asphalt of Fifty-fourth Street, they saw the marks of survey crews. Pink spray-painted lines showed the new highway's route, heading directly into the field. The real world— Babylon, they sometimes called it—was getting closer.

Whatever the activists did seemed to make no difference. Once, earlier in the summer, they had danced and drummed and chanted in the echoing rotunda of the State Capitol. Natalia and three others stepped over the velvet ropes protecting an inlaid brass star, the "Star of the North" of Minnesota's state motto, gleaming on the marble floor. They sat down, U-locked their necks together, and raised their fists.

Natalia (center) and three other demonstrators lock themselves together by the necks in the State Capitol Rotunda. Staff photograph by *St. Paul Pioneer Press*; copyright *St. Paul Pioneer Press*.

"It was like, 'Oh my God, they're on the sacred seal of the great state of Minnesota!'" Natalia said mockingly. "Which we thought was really funny. We were like, 'OK, if you're going to fuck with our sacred sites, we'll fuck with your sacred sites. You want us to get off your sacred seal, get off our sacred site.'"

The official response was not overwhelming. "They're entitled to protest—certainly they are," Governor Jesse Ventura told a newspaper reporter, "but to me it's a done issue. They tore up that street when I went to high school and no one was protesting back then—where were they? They show up in the last two years, but that's certainly their prerogative."

One Capitol worker was even less impressed. "They *smelled*," he said, appalled. "They were so *dirty*. I had no idea body odor could be so *pungent*." It was unbelievable, he said, how the smell had filled the rotunda. He wondered about rumors that protesters at the camp were getting paid. "Professional protesters," critics sometimes called them. The idea would have seemed strange to the Earth First!ers, subsisting as they were on food brought by supporters, free leftovers given out after hours at the Seward Community Café on Franklin Avenue, and vegetables scavenged from Dumpsters.

As summer faded into fall, the activists' dogged guerrilla war dragged on. All night, sentries stood watch along Hiawatha Avenue. Bicycles flitted through the shadows cast by the lights at the gas station and used-car lot. Often, the chain saw crews struck in the early morning; they would seal off an area, cut the trees, and leave them lying. Each time the crews appeared, the sentries would radio frantically for help. Since the movement didn't believe in leaders, the response from camp would depend on whoever was on hand and what they decided to do at that moment. Asked later what the overall strategy had been, Natalia summed it up this way: "We learn that they're there, and we run."

"At that point, we were just trying to be all places at all times, and if we weren't, we'd haul ass. It was all out," said Dr. Toxic.

"It was a very reactionary, military situation," Wes, the self-mocking, long-ago Santa Claus, remembered. And he could see that they were losing more ground every day.

Then in late September, a power line worker passed somebody in camp a tip: the big cottonwood overhanging Minnehaha Creek was to be cut that Monday. It was time to try something new, the activists thought.

"We needed to be aggressive and escalate the conflict, instead of being reactionary and defensive," Wes explained. "We were in a position of warfare, you know, where imperialism is encroaching and destroying and destroying and destroying, and leaving nothing, and we can't really win. It's just a matter of time. It's just like, you can look on the calendar and about guess when you're going to be wiped out entirely."

So somebody—not Wes, just some people who were around and energized that day—got the idea of bringing out the old steel tripod, the one to which Joe Hill had chained himself when the protest first began. The plan was to have it sitting on the highway at dawn on Monday, with an activist chained to one of the tripod's steel legs. That way, they figured, the crews couldn't get their truck through to take down the cottonwood.

For months, people at camp had been taking shifts in the cottonwood tree. When Monday came, curly-haired Freedom, now graduated from high school and free from homework, and her friend Rory, nineteen, had been up there for about a week. Mornings, lying in their hammock among the yellow leaves, they watched the sun come up and listened as the stream of traffic on the highway swelled with commuters, roaring by just a few feet away from the tree's trunk. They wondered where the people were going. Look at this road, Rory said to Freedom. It just doesn't end. It just keeps going and it doesn't go anywhere. Western civilization was like a virus spreading over the face of the planet, Rory thought. Fighting it was like being a white blood cell.

On Monday morning, Rory and Freedom woke to a clanging. Below them on the highway, they saw somebody setting up the tripod. "It was like, *yeah*!! Take it back! Take it *back*!" Rory said later. "Tear up that asphalt right now."

When the police came, they shut down the entire highway. The traffic backed up for blocks. It was late afternoon before they cut the protesters from the tripod. All day long, lines of detoured, angry motorists wound through the neighborhood's back streets. As the sun went down, three squad cars sat underneath the cottonwood, enclosed by a line of yellow police tape. Wes, Dr. Toxic, Squash, Nettle, and a sixteen-year-old runaway called Midnight had joined Rory and Freedom up in the branches. Now, Midnight dangled uncertainly from a line strung between the cottonwood and a nearby oak. A crowd of around seventy-five people watched from the grass below.

"Cut the rope! Cut the tree down!" a man yelled. "Get 'em out of the neighborhood! They don't even live here!"

"They're very selfish people, holding up people who want to get to work and get their kids," said a woman in a sweat suit.

"We want a wide road, the wider the better!" somebody yelled.

"What I want to know is when they had the raid in December, why they didn't have twenty guys with chain saws cut down these trees and nip this in the bud." another man said.

"I can tell you all my neighbors support this reroute. I've been to meetings, and I can tell you everyone accepted it. You can just listen to the neighborhood talk. They want 'em out of here," someone said.

"All these idiots belong in a mental institution," muttered somebody else.

"Bag 'em up and take 'em away," said a guy in sunglasses.

Other onlookers were more sympathetic to the protesters. "I'm not saying they're wrong," a man said. "I'm just saying they're going to lose."

The crowd grew. A cherry picker pulled up next to the cottonwood. A Minnesota State Patrol trooper and a worker with a chain saw stepped into the bucket at the end of the cherry picker's long yellow arm, soared through the air, and hovered near the tree. The trooper pointed his baton at the branches. The chain saw whined. Limbs began falling with a scatter of golden leaves. Rory and Freedom and the others scurried like treed raccoons, dark shapes in the fading light.

A thick limb dropped heavily, and the whole tree swayed. It looked for a minute as though Midnight might fall. "He gave the ultimate sacrifice. He went down with the tree," a man in a polo shirt commented sarcastically.

"Please don't cut the trees!" screeched a woman with a bone in her nose.

"You can scream all you want and they ain't going to change, lady," said a middle-aged black man standing nearby. He lived farther north along the highway, where the crews had already come through. "At least when they cut ours down they snuck up on us. I remember when I came home I almost cried," he said. "Now everybody coming home from Burger King, they shine their lights right in my window."

As the chain saw lopped off more and more of the tree, the moonscape of the road project behind it came into view: caterpillar tractors crawling over banks of naked dirt. Behind, the skyscrapers of downtown Minneapolis glinted in pink light. Firefighters pulled a hook and ladder truck beneath the cottonwood and began cutting down hammocks. A fireman sawed at a rope that held sacks of canned goods and water.

Rory leaped from branch to branch. Suddenly, as the crowd below watched, he dropped onto the fire truck's moving ladder. His wild hair, baggy pants, and big boots silhouetted against the sunset, he did a frantic dance, then reached down with a bicycle U-lock and froze the ladder's rungs.

Later, Rory didn't remember being at all afraid of falling. He felt as though invisible hands had lifted him up, the hands of all his friends at the Free State. Only a few months earlier, he had been living in a tent in a city park after his mother kicked him out of the house. At the Free State, it seemed to him, he was part of a real community for the first time in his life.

Police pulled Rory to the ground and he landed on his back on a cement sidewalk, but he wasn't hurt. He felt as though he had been protected somehow. "Having all the people around me and being up there with all my friends, it felt, it really felt like something beyond me," he said later.

Wes began cutting himself once on the chest for every limb the chain saw took. Finally he leaped onto the arm of the cherry picker and was arrested. Freedom went down next.

By nightfall, police spotlights swept the four who remained in the cottonwood tree and the crowd who watched from the dark, including the tiny grandmotherly figure of Carol Kratz. "I still hope it can be stopped," she said. "I pray every day." The chain saw droned. Flecks of sawdust caught by the spotlights fluttered down like snow.

Hours later, after the cherry picker and the fire truck had left, Dr. Toxic climbed high in the ruined cottonwood's remaining branches. He peered through the blackness at the crowd below.

"Whose park?" he yelled down.

"OUR PARK!" chanted supporters.

Rory dangles from a line as he and four others in a cottonwood tree try to stop workers from cutting off the tree's limbs. Photograph copyright 2001 *Star Tribune*/Minneapolis–St. Paul.

"Whose trees?"

"OUR TREES!"

A few days later, only Nettle was left in the cottonwood. There weren't enough blankets to keep four people warm at night, and there was only so much food. So one by one, Midnight, Dr. Toxic, and Squash had climbed down. Nettle was the one who most wanted to stay.

Nettle had been a wanderer for a long time: ever since she was twenty, as near as she could remember, so that made seven years on the road. She had spent the last three winters at Big Mountain supporting the Navajo elders. Before that, she had done ground support for Julia Butterfly Hill's vigil in the redwood. For a while, Nettle lived on some land there in California, planted a garden, and prayed with Native people for the forest. Once, long ago when she was still in college, she had stayed with Mayan people in Belize.

Nettle had a soft voice, a gentle, dreamy manner, and brown hair that fell in dreadlocks past her shoulders. On her wrist she wore a moon and star pattern she designed herself, her first tattoo. Later, after Squash learned how to do tattoos with a pointed twig, Nettle got him to add a Celtic love knot on the inside of her arm, and a design from the *I Ching* on her index finger. Nettle believed in helping indigenous people in times of need. That was why she had come to Minneapolis, and why she was defending the cottonwood. Dakota people considered cottonwoods the tree of life, she explained later. They used cottonwoods in their Sundances and sometimes saw visions of their ancestors up in the branches.

"I felt like it was such an honor to be up there in the cottonwood tree," Nettle said. "It felt like a grandmother tree to me. By the seventh or eighth day I felt like I was having conversations with the tree, and that she was comforting me. At one point she told me that all these people down there supporting us on the ground and all the people up in the tree were her grandchildren. The little trees, they'd all been cut down, they weren't there now, but she said that all these people who were standing up around her, that we were like her grandchildren, and that she wanted to share her wisdom and her strength with us."

A few more days passed. September turned to October, and night temperatures dropped near freezing, but Nettle hung on. "It was real cold, and I would just wake up, and all I'd have would be like one can of almost frozen food," she said. "I'd be shivering and I'd have about one or two sips of water a day. It was challenging. It was a lot different from being out in the redwoods."

At the base of the tree's big trunk, squad cars sat night and day to prevent anyone from bringing Nettle food or water. The traffic on the highway never seemed to stop. Some times passersby yelled. "They would say, 'Shoot 'em down with water! Get a fire hose out there and spray 'em down!'" she recalled. One time a masked man showed up carrying a sign saying "On With Progress."

But other times, motorists honked in support, and every day people from camp stood on the footbridge near the cottonwood. "I would start to have a hard time, and there was always some-body out there," she said, "just keeping me strong, keeping the tree strong." A couple of days in a row, a school bus slowed way down as it passed the cottonwood, and children inside yelled, "Go Nettle! Save the park! Save the trees!"

"I got such a burst of energy," she said, "that all these little kids knew what was going on, that they knew the right thing."

But every day she was up there things got scarier. The mangled cottonwood was lopsided; Nettle could feel it starting to lean toward the creek. One day construction crews started work-ing around the tree's base. "It felt like an earthquake," Nettle recalled. "I thought for sure at that point that the tree was going to topple over and that I was going into the river."

At camp, Nettle's friends were getting worried. Squash said Nettle needed to be relieved. She was a strong woman, he said, the strongest he'd ever met, but she was getting tired. Dr. Toxic, before he climbed down from the cottonwood, had promised Nettle that the camp would send her supplies and relief. They just had to figure out how.

Dr. Toxic had a pair of what he called "bark walkers": spikes that loggers strap to their boots so they can walk up trees. Now he and Freedom spent a whole day practicing climbing with the bark walkers. Using his old army training, Dr. Toxic also taught Freedom how to crawl through the brush on her belly without being seen.

That night, wearing the bark walkers, a camouflage suit, and a backpack of food, Freedom wriggled along the creek toward the cottonwood. Out on the highway, Dr. Toxic distracted the cops stationed at the big tree's base. Freedom hurtled onto the trunk and scrabbled upward, but not fast enough. The police arrested her and confiscated the bark walkers. Soon after that, officers moved the barriers farther away from the tree, so no one could get close to it. "So that's when we were really desperate," Dr. Toxic said. "That's why we ended up doing this huge, crazy, I-can't-even-believe-we-got-as-far-as-we-did plan."

The plan went like this. This guy named Treefrog would climb a telephone pole, shoot a length of lead-weighted fishing line across the highway to the cottonwood with a slingshot, then tie on a rope. Nettle would pull it across, and Treefrog would slide down the rope into the cottonwood tree with a pack of supplies. Treefrog would take over for Nettle. He would even have a filter so he could dip his drinking water out of the creek with a bucket and a rope. He would be able to stay up there for ages.

"That was just the most amazing plan," Dr. Toxic recalled fondly. "In fact that was like the plan before we had Freedom go up. A couple of different people had mentioned, let's try the telephone pole thing. But we were like, that's just too James Bond. Let's try something more realistic first." He laughed, remembering. Even for a cop-defying, skyscraper-rappelling guy like Dr. Toxic, it was an outstanding plan.

With the barriers moved so far away from the cottonwood trunk, Nettle felt more cut off than ever from her friends in camp. Now when they shouted to her, she couldn't hear them. She had been up in the tree for nine days. She thought maybe the police would be coming for her soon, and that would mean the end of the cottonwood tree. "I felt like I was next to the deathbed of an elder or something," she said. "But the tree told me that it knew that we had tried our hardest, and that it knew that it had to go, but that I would see it again. It said it wouldn't be in this physical world again soon, but it would still be with us. And so it told me that I had done all I could and that whatever happened it was time to go down."

That night Nettle looked out from the branches and saw people from the camp gathering outside the barricades for a candlelight vigil. To her left, Native people gathered with their drum, "a real strong presence of indigenous people all standing together," she said. "Everyone had candles all around. It was starting to get dark, and over to the right, all these people were singing, just like amazing, beautiful, really in-harmony songs to the earth. So it was just this beautiful thing."

Below her a car pulled off the highway, and then she saw Treefrog climbing a telephone pole. A fishing line arced across the highway, but fell short. Treefrog tried again and again, but each time the line failed to reach quite far enough. A truck with a cherry picker pulled up the telephone pole, and troopers hauled

Treefrog in, lowered him to earth, and stuffed him into a squad car. Then the cherry picker pulled up to the cottonwood.

Nettle climbed higher into the branches. Soon the officers were cutting what was left of the hammock along with the safety lines that kept her from falling. They told her a storm was moving in. Work with us, they said. She felt wobbly and nervous and lightheaded from hunger. She had done all she could. She climbed into the cherry picker and rode down.

Later, when people from camp retold the story, they called it the Siege of the Cottonwood. They too had done everything they could, and it wasn't enough. "That was when I started looking beyond the Free State and saying this is going to be destroyed, what are we going to do about it?" Wes said. "What do we want to save out of this whole experience?"

Wes didn't like to talk about his role in the siege, especially how he had cut his chest with the knife that first day, as the tree shook and the limbs and leaves fell around him. When he was asked about it, his self-mocking manner fell away for moment. "I wasn't scared at all. It wasn't scary, you know. It was just a vision of the apocalypse," he said, his voice flat.

Wouldn't a vision of the apocalypse be scary?

"Not for me personally," Wes replied. "But for the world, yeah."

By late October, the bridal wreath hedge along what was once Carol Kratz's driveway turned a brilliant orange. The broccoli in the garden had gone to seed, half buried under fallen leaves. All around it, big oaks that had once shaded Riverview Road lay dying. The crews had come without warning, while the cottonwood siege was still on. They had left some trees for now, and they hadn't reached the camp at the end of Riverview Road yet, but the activists all knew the crews would be back.

As soon as they heard about the cutting on Riverview Road, Jim Anderson, Marshall Law, and the red-haired, middle-aged activist Susu Jeffrey went to inspect the damage. "It's like a drive-by shooting," said Susu. She went to get Carol Kratz at her new house a few blocks away. Marshall called a television crew. Carol wept in front of the camera. "You used to live here?" asked the reporter. Carol said they took her house, her husband had died, and now they took her tree. It was all true, but somehow it fell flat. She couldn't pull it off for the camera.

The television crew left. Jim burned some sage, and they sat on a downed log amid the wreckage. Jim looked old and tired. Wes came walking up the street barefoot, his dark-rimmed glasses now held together with duct tape. He took a marigold from behind his ear and gave it to Carol.

After a while, she went over to the big oak that once sheltered her house and put her hand on the bark of its now-horizontal trunk. A tiny figure in a pink sweat suit in the midst of the wilting leaves, she looked much sadder than she had a few moments before, when the camera was still rolling.

Along Hiawatha Avenue, the new highway was taking shape. The patch of razed parkland by the Longfellow Lagoon was buried under mounds of dirt. Beside it ran a ribbon of new asphalt, a temporary bypass that would keep traffic moving while Hiawatha Avenue was widened. Bulldozers had leveled a wide, flat road-bed. It was possible now for anyone to picture the new, widened highway the engineers had seen so clearly. It would look, of course, exactly like the bleak, generic landscapes that crisscross the country along highways coast to coast. The only thing hard to imagine from now on would be what it had looked like before.

To the south in the field by the four oaks, on a grass-covered mound that served as a lookout post, Marshall Law stood sentry, a stocking cap pulled down over curly hair of an indeterminate shade, neither red nor brown nor blond. He was watching for chain saws, he joked. Helicopters. Humvees. He had long pale eyelashes, an intense gaze, and an intelligent, ironic manner. He would be thirty in November, he said, his voice incredulous. "Winter's coming on again," he said. "I'm really wondering where the time went. Fourteen months is a long time." He paused. "I think a lot of folks are really wondering where we can go from here."

A few blocks from the lookout hill, the Hydro-Ax had been eating its way through the trees and brush behind the used-car lot and some aging condominiums that fronted the highway next to the corner gas station. It was a warm October afternoon, and somebody from camp had locked himself down to the big machine and been taken on a wild ride to a nearby highway garage, where

Tree and brush clearing in preparation for the Highway 55 reroute. Photograph by Peter Hoseth.

he was cut loose and arrested. Now the Hydro-Ax was gone, but it would be back. "Every time we do this they're getting closer," said a middle-aged Native American man called Thunder. He had grown up on Forty-eighth Avenue, a few blocks away. "This area they just cut, that's where we used to hang out when we were kids, build forts and stuff," he said.

Marshall Law and a couple of others sat in a stand of sugar maples nearby. "As far as what trees are on the front lines, I think these are the most important," said Dr. Toxic.

"I think we're probably safe for today," said Jim. "They know we've activated too many people now."

A sliding glass door opened at the back of one of the condominiums, and a portly middle-aged man in a black and white bowling shirt stepped out to talk to Jim. The man had a Motorola radio strapped to his shirt pocket. People in camp knew him by the name Rattlesnake. He had been their friend for months. He had given them CB radios and helped them with their communications system.

Once the highway went through, the picture windows of the condos where Rattlesnake lived would look out not on woods, but on a story-high concrete sound wall. "I've been against this for a long time," he said.

"I support them [the protesters] when I can," Rattlesnake said. "I don't have the time and energy and guts they do. I kind of feel they're fighting my battles for me, and I appreciate it very much."

By late October, the Hydro-Ax had finished shredding the trees and brush behind the condos and moved on to other pastures, but nearby, the stand of sugar maples was still intact. Some of the trunks had brown, red, and blue yarn netted around them, like a protective spell. A green construction-paper sign had been taped to another: "Warning. Trees have been Spiked. ELF."

It wasn't ELF's first appearance. Earlier that month, someone had damaged equipment belonging to C. S. McCrossan Inc.,
the construction firm building the highway. The company
declined to divulge the extent of damage, but Bob Greenberg
e-mailed news organizations claiming that hoses and wires had
been slashed and sand had been poured into oil reserves. The
e-mail, he said, was a communiqué he had forwarded from ELF.

ELF stands for Earth Liberation Front, a hybrid of radical
environmentalists and the Animal Liberation Front, or ALF, the
radical wing of the animal rights movement. In 1998 ELF claimed
responsibility for fires at a Colorado ski resort that caused more
than twelve million dollars in damages. It was not at all clear
whether the people behind the resort fire were the same ones at
work in Minnesota. Whoever was responsible, the damage done
was orders of magnitude smaller. Still, a faint whiff of ecosabotage
hung in the air. *Warning. Trees have been spiked. ELF.*

On the back wall of the condos, somebody had hung a huge
hand-lettered banner: "MnDOT GOONS, don't rape the land."
James McNamara, a thirty-three-year-old freelance opera director
who lived in the condos, stepped out onto the lawn. McNamara
said he hadn't seen any evidence of tree spiking, but thought he
heard some. "One night I heard this tap, tap, tapping all night
long, and I knew it was the Earth First!ers," he said. "In a weird
sort of way, it made me feel safe."

A month later, the sugar maples still stood, just outside
the path of the highway. Many other trees in the area had not
survived. Clearing for the widening of the existing highway was
nearly finished, from the now-demolished thickets by the Longfellow Lagoon all the way south to the camp's border at Fifty-
fourth Street, eight blocks in all. On the section of Hiawatha
Avenue that skirted the park, the stately shade trees were now

stumps, decorated here and there with small offerings: a pile of stones, a jar of water, a blue jay's wing.

At Riverview Road, near where Carol Kratz's house once stood, oak logs two or three feet in diameter were piled high. The air was fragrant with the sharp, sweet smell of freshly cut wood. Tags on the logs indicated that MnDOT was sending them to an organization dedicated to the preservation of Dutch fishing boats.

Most of the river side of Riverview Road was a cleared field now, waiting for the earthmovers. Only the trees closest to the steep gorge of the Mississippi remained. It was as if the landscape had suddenly shifted and the ranks of oaks had retreated to the safety of the river.

At the southernmost end of Riverview Road, near the camp border, a few big oaks still stood. On a sunny day with temperatures in the fifties, mild for Minnesota in November, people from camp occupied every tree. More campers lay around on the dry grass underneath, relaxing in the warmth.

Caleb, the intense, haunted young man who had stood by the fire with Bear that bleak winter day, was cheerful today. He told a story, illustrating it with wide sweeping hand movements and much bounding around. People laughed as they do with a well-loved member of the family.

Caleb was well liked at camp. Often he was the first friend that newcomers made. He could play the guitar and make up amazing lyrics off the top of his head. He said odd things, but sometimes they were just what was needed to keep everyone going. People from the Free State knew Caleb was different, but they bristled at the suggestion that anything was "wrong" with him. "Caleb is not crazy," is how Marshall Law put it once. "America is crazy. He kept us sane in this insane world."

Now a reporter asked Caleb what he would do when the camp no longer existed. Caleb paused, and two women, his friends, said, "You're going down the Mississippi."

Caleb looked at them. "Am I building a boat? Or are you?"

"We'll float the van down, remember?" they joked. "We'll float the biffy down."

Nearby, a man named Paul Eaves stood in the sun. He was casually but conventionally dressed, with a neatly trimmed gray beard and short, wiry hair that stood on end around his face. He lived two miles away and was a "part-time occupier," he said.

Paul Eaves was also a witch. He belonged to a coven called 100 Witches. (Modern male witches do not call themselves warlocks. It's a term they associate with male informers who they say infiltrated covens back in the days when witches were burned at the stake.) The faith, known as Wicca, that the coven followed was an "authentic nature-based European tradition that's been lost," Eaves explained. Many people were called to this religion, he said.

Several years ago, when he was out looking for an open private place where the coven could hold ceremonies, he had discovered the four bur oak trees. "I just stood in the midst of them and said, 'Oh my, there is a particular sort of energy in that space,'" he said. The coven had held several maypole ceremonies and other rituals under the four trees. They had also built the wood chip labyrinth nearby for "walking meditation."

Once, Druids had danced in sacred groves, but now the connection between people and the land had been severed, Eaves believed. The Highway 55 project was a perfect example. "It's emblematic of a lot of things that are wrong with the culture right now."

Marshall Law looked down from an oak a few trees over. Asked if a visitor could climb up, Marshall descended gingerly to the bottom branches to toss down a rope ladder, feeling cautiously with his feet. It had been hard, Marshall said, watching what used to be Riverview Road disappear. "I walked through here in the middle of the night and had the oddest sensation of not knowing where I was," he said. Now the people in the trees were just trying to hold out for a little while longer, he said. They had only so much food and warm clothing, and it was getting cold. "The wind does blow a lot stronger fifty feet above the ground," he said. "We're telling people we're just trying to buy a little more time for the Mendota."

The Mendota Dakota were trying to negotiate something, Marshall explained. There was a meeting set up for December 2 with Minnesota senator Paul Wellstone, he said. There might be congressional hearings, or federal intervention. "At this point, I think that's the only place that hope lies," Marshall said.

Marshall Law in the branches of an oak tree at the edge of the second camp. Photograph courtesy of Mendota Mdewakanton Dakota Community.

(Asked later if there really was a meeting set up with Wellstone, Linda Brown said, "Sort of." Somebody had requested one. They were waiting to hear about it.)

A young Native American woman called Soil walked over the dry grass to the base of the tree, carrying a paper bag with a plate of lentils and sweet potatoes. Marshall lowered a rope and brought it up.

To the north, the Hydro-Ax gnawed up stumps. It reminded Marshall of the contraption that consumes all the truffula trees in the Dr. Seuss book *The Lorax*. The Hydro-Ax sprayed wood chips with a force strong enough to pierce the plastic water bottles hanging from trees. When it came too close, Marshall explained, you couldn't look at it without getting blasted. You had to get on other side of the trunk.

The Hydro-Ax lumbered closer to Marshall's tree, then sprayed a cloud of dirt and wood chips high into the branches. Marshall moved to the other side of the trunk and held on.

CHAPTER 9

Four Bur Oaks

Minneapolis police lieutenant Bud Emerson would be glad when the whole thing was over. Privately, he was sorry to see so many trees lost in his neighborhood. It had been hard to watch, all that summer, but he had a job to do, and that was to uphold the law and make sure nobody got hurt, including the protesters.

"They scared the bejesus out of the construction guys," Emerson said. "You're trying to operate a backhoe and some guy suddenly runs up and chains his neck to the operating part of your equipment, that's a little scary." No construction worker wanted to play a part in killing someone. Emerson was afraid a worker would suddenly swerve with a road grader: "He could tip the fool thing over and get himself hurt." He also worried about the protesters with their signs out by the highway: "We were concerned about them deciding that they were going to do a political action that included jumping in front of traffic and getting creamed."

Emerson wondered sometimes about the psychology of the protesters. They talked as though the police and the workers were evil incarnate, he had noticed, yet they were constantly entrusting their lives to the forbearance of the construction crews. "If you really thought those people are totally evil, why would you run up and chain your neck to the front bumper of a thirty-thousand-pound earthmover?" he asked.

Sometimes the protesters would stand just outside the yellow police tape that surrounded the construction site and stare. It reminded Emerson of a schoolyard turf war. "Somebody runs over to your turf, puts their toe over the line, says 'Nyeh nyeh nyeh nyeh nyeh,' and runs like hell," he said. "I told my people, 'Monitor the situation. Don't chase people down the block.'"

What had to be done was pretty straightforward, as Emerson saw it: just whittle down the trees until none were left. "Basically we would just tell the cops to kind of keep an eye on them and let us know when there's nobody in the tree," Emerson said. Then police would tip off the crews. "They would kind of at random sneak down there, run a truck up to the tree, cut it down, then get out of there before the protesters could surround it."

The cottonwood occupation had been a knottier problem, Emerson recalled. He and the other officers worried that protesters in the branches might drop down onto trucks entering and leaving the construction site. And the spectacle in the cottonwood was a traffic hazard, distracting motorists just as they entered the construction zone. When the protesters set up the tripod on the highway, police decided it was the last straw. Emerson and other officers decided the tree was "a center for provocation," he said. It was time to get the protesters out of it.

So Emerson and the other officers sat down and deliberated about how to end the cottonwood problem with the least risk of hurting anybody. They decided to cut off the tree's lower branches, figuring the protesters would retreat upward till they had so little tree left they would have to give up.

At first the plan had seemed to work, but then the protesters figured out what was happening and started challenging the guy with the chain saw. "It got a little dicey," Emerson said. He watched from below as the protesters leaped from branch to branch, trying to get ahead of the moving bucket of the cherry picker. They were getting frantic, he thought, when one guy leaped onto a ladder and another clung to the cherry picker's arm. He wasn't sure what their exact mission had been.

Emerson had joined the police force because he wanted to make a difference in the world. It bothered him when the protesters asked police how they could do what they did and still be human beings. "I don't see myself as a bad guy," Emerson said later. "In any society where there's free expression, everybody can disagree but at some point somebody has to make decisions about what's the right thing to do, and the rest of us have to stand by and let that thing happen."

Still, in this case standing by and watching was hard. When his two grown daughters were small, he had walked with them near this same doomed cottonwood and they had played on the footbridge in its shade. He and his wife had often taken the little girls for picnics in the now-razed parkland by the Longfellow Lagoon. "Throughout this whole thing, there was a sense of sadness on my part that a piece of my family's history was being destroyed," Emerson said.

In the fall, his family had enjoyed walking to the lagoon and collecting chestnuts from a tree that had somehow survived the blight that wiped out almost all of the country's American chestnuts decades ago. The little girls had called the place "ga ga land," for some reason he didn't remember. In the winter, the family had skated on the lagoon, which in those days the park board maintained as a skating rink. "It was like a picture out of Currier and Ives," he said. "They'd have lights up in the night. They'd have this long serpentine area that they would plow out and spray down so you could go out there and skate. It was beautiful."

He had to stand there and watch as the place was razed. It

hadn't helped when afterward one of his neighbors had been so angry about it that she wouldn't speak to him.

In November 1999, Bud Emerson drove out to the camp, accompanied by Captain Kevin Kittridge of the State Patrol. They stood with Jim Anderson, Bear, Thunder, and an Ojibwe elder, Darlene Jackson, in a circle underneath the four bur oaks and passed a red pipestone pipe. The officers had come to talk about clearing the camp. Both sides knew a raid was coming soon, although the officers wouldn't say when. The officers did say they wanted this next raid to be different from the last.

Kittridge was a big beefy man whose uniform was tight at the neck and armpits. He looked like a trooper out of central casting, and he spoke carefully, with a trace of the accent parodied in the movie *Fargo*. But he was not a stereotypical, take-no-nonsense-from-anybody kind of cop. He was willing to listen to what the protesters had to say. All through the summer and into that fall, he had been going out to the camp to meet with Jim and Bear.

Kittridge saw smoking the ceremonial pipe as a way of convincing Jim and Bear that he could be trusted. "What Jimmy always called the pipe, he called it the truth pipe: 'We're gonna smoke the truth pipe.' And I didn't know what that meant, but I assumed that it was a bond, that you gave your word," Kittridge said. The meetings were tense at first, but less so as time went on. "I think we all agreed that at the end the only thing that any of us had was our word. There was nothing else. I think once we figured out that everybody was going to be honorable, things went better."

Kittridge acknowledged that in the first raid, back in December, the officers had destroyed drums and slashed tipis. "I think that when we came the first time we had no sense of, in the Native American community, what was sacred to them, what was important to them," he said. "I think that a lot of the things that they held sacred, like the feathers and the drums, got trampled as we rushed in."

After dawn on the morning of that first raid, Kittridge offered to help Jim save the sweat lodge behind the houses on Riverview Road. (In the daylight Jim could see that the sweat lodge hadn't burned after all.) "He was explaining to me that the sweat lodge was no different from my church," Kittridge said. "So I took one of our SWAT teams and we went across and we took this thing apart, while they instructed us from across the street: cut this, don't cut that. We took it apart, rolled it all up, took it back across the street, and gave it back to them, and they stuck it in Bob's van and took it home with them. I think maybe it started us down the road of, let's try and understand what's going on here, and what we can do to make the best of a bad situation.... As I told my troopers all the time, the bottom line is if you say, 'Win, lose,' we're going to win. They're going to lose. At least we can let them walk away with some dignity."

So that November, nearly a year after the first raid, Kittridge and Emerson and Jim and Bear negotiated some ground rules for the second one. The officers agreed that they would move in only after the sun was up; they would allow time for a ceremony around the four trees, then give everyone a chance to leave. Those who chose to stay would be arrested without pain compliance holds or chemical sprays. Workers would dismantle the camp under the supervision of Jim, Bear, Thunder, and Bob Brown, taking care that ceremonial objects were handled with respect.

In return, Jim and Bear agreed to a list of the officers' requests. No one would booby-trap the woods; they would agree not to occupy a dugout in the side of the lookout hill or the disused twelve-foot-diameter water main that ran from the VA Medical Center under the camp to the Mississippi.

The Native Americans did not intend to get arrested, Jim told Kittridge. They would stay by the four trees and pray till it was time to leave. But Jim and Bear guaranteed that the Earth First!ers wouldn't make the police run through the woods or hurt their backs carrying arrested protesters over rough terrain. Kittridge figured Jim and Bear would make the agreement stick.

"That was the big difference for us between their group and the Earth First!ers," Kittridge said. "When I'd go down there and the group would come together, the Earth First!ers would start with the rhetoric. And eventually Jimmy or Bear would just tell them, 'Shut up. Let's talk business here.'" Bear reminded Kittridge of a summer camp counselor making kids behave.

"I found him to be real honorable," Kittridge said later of Jim Anderson. "He was real emotional about this, he thought what we were doing was very, very wrong. But when we'd come to

Tipis and sweat lodge near the four bur oak trees. The VA Medical Center complex sits on the low hill in the background, known to the Dakota in the 1800s as Taku wakan tipi, the dwelling place of the gods. Photograph courtesy of Mendota Mdewakanton Dakota Community.

an agreement about who I was going to bring into camp, or what we were going to do once we got in there, I never in all my dealings with him saw him do anything other than what we'd agreed on. And Bear was the same way."

In early December, the Mendota Dakota met with Minnesota transportation commissioner Elwyn Tinklenberg to make a final, desperate request. They wanted the road detoured to save the four bur oaks they so passionately believed were sacred. Maybe the oaks could be left standing in the median strip, they thought. Somebody from MnDOT had once made the offer, Bob Brown seemed to remember, way back at the beginning of the year during the federal mediation. At the time, Bob had been so sure they would stop the road completely that he had dismissed the idea. "I said hell no. They asked me what if the road was twenty-five feet from the trees, or fifty feet from the trees. We said we can't have that—traffic whizzing by when you're trying to have ceremonies." But now, with the end coming closer, he thought maybe the offer could be reopened.

It was too late, of course, as anyone more familiar with the process would have known. At the meeting, Commissioner Tinklenberg explained that the engineering work was done and the construction contracts already let.

Tinklenberg, a soft-spoken former Methodist minister, seemed troubled that there was no way to build the road and still accommodate the Mendota Dakota. If there were, he told reporters, no one would be happier than he would be. He said MnDOT would look back through the records and see if an offer to detour the road was ever made.

As it turned out, no one had kept minutes of the mediation sessions. The Mendota Dakota hadn't thought to ask for them. It seemed to Bob Brown that Mike Haney had taken some notes—

but Haney had disappeared months earlier, and the Mendota Dakota hadn't heard from him since.

That night after the meeting with Tinklenberg, Bob and Linda Brown drove out to camp in the dark. The oak branches stood bare against the backlit city sky. They could hear the drums a block away.

Jim, Bear, and Clyde Bellecourt, the aging cofounder of the American Indian Movement, stood under the four trees, by the flickering light from the cauldron that held the ceremonial fire. On a blanket on the ground nearby, among sage and cedar branches and beaded leather pipe bags, lay a stack of wrapped sterile surgical scissors and razor blades. Someone held up a lantern, illuminating the face of the gray-bearded, wiry-haired Wiccan Paul Eaves. He held a ceremonial pipe, the bowl toward his bare chest, the stem toward the sky. Clyde pinched the skin of Paul's shoulder, then cut off a small chunk of flesh. The two men embraced, and Paul went to join a line of bare-chested men with bandaged shoulders near the fire. Jim took the flesh, wrapped it in colored cloth, and tied it with sinew into a bundle.

In the shadows beyond the lantern light, men and women waited their turns, as if for some strange vaccination. Solstice stepped forward, then Bob Brown's younger sister, Linda M. Brown. She pulled down the neck of her sweatshirt, baring her shoulder and a white bra strap.

Emily stepped from the shadows and explained to an on-looker that the ceremony was a "flesh offering." Harry Charger, the Lakota Sundance chief from South Dakota, had given Jim Anderson the instructions. They had come to Harry Charger in a vision.

Emily and more than fifty others had already gone through the line. "It meant that I'm willing to give blood to protect this place. It's a symbol of how much people are willing to sacrifice," she said. "I hope this will be the last blood to protect this area." Remembering, perhaps, the time police banged her head on the railroad tie, she added, "I don't want anybody else beat up by police officers. I don't want anybody else hurt."

The pile of sinew-bound cloth bundles grew. Jim told people to tie them to the four oaks, with red to the west, white to the north, yellow to the east, and black to the south. Others wrapped the oaks' trunks with ceremonial cloths and moved the cauldron out from under the oaks' branches and into the field, as Harry Charger had instructed them to do. Then the camp's supporters melted into the night and drove home under the harsh bright lights on Hiawatha Avenue. Behind them the field was quiet and dark except for the glowing fire.

Ten days later, in the early morning of December 11, activists' phones rang all over the Twin Cities. In the field by the four oaks, the drums began to beat—*LOUD soft*, *LOUD soft*, *LOUD soft*. Cars and pickups bumped across the snowless field, parked in a circle, and shone their headlights on the four trunks.

A pipe of red pipestone carved with an oak-leaf design sat inside a circle of sage on the ground in the center of the four trees. Harry Charger had told Jim to put the pipe there and leave it. Anyone who picked it up now, Charger said, would bring misfortune on himself and generations of his family.

Close by, a pale tipi stood against the sky. High clouds passed over. One fuzzy star seemed caught in the tipi's poles. The tipi seemed for a moment as though it stood on the plains a hundred years ago. In the dark beneath the four oaks, a man sang to the beat of a single drum.

People clustered around the cauldron, holding their hands out to the warmth. Others sat in the hay bale kitchen. As it grew colder, those outside in the dark hunched their shoulders against the wind and drifted toward the camp's winter supply of fire-wood, piled in a big heap. "Torch it," Jim said. "They're going to take and mow everything over anyway. We might as well use it instead of them."

Someone lit a match beneath the wood. The fire caught and roared. A crowd of about fifty stood illuminated by the flames. They wore sweatshirts and stocking caps and blue jeans and battered hiking boots. They wore beards and shoulder-length hair, or long flowered skirts, or overalls. Some had blankets around their shoulders, like refugees. They looked like people who hung out at alternative coffeehouses, lentil eaters, readers of left-wing literature: a crowd police could take one look at and know right away that they weren't ordinary Minnesotans, that these weren't people like them.

Flames shot ten feet into the air; sparks showered down. The fire must have been visible a long way off.

"OK, listen up," said Bear. It would be eight more minutes till the police came, he said. "We want everybody to know we're praying for you. No matter what happens, we're still going to be a family. We love you all for what you're doing," he said to the young Earth First!ers. "We're going to stand by you one hundred percent. *Mitakowasin*," he ended, the Dakota word meaning "all my relatives."

Jim Anderson spoke next. "What they're doing to us is nothing we have to feel bad about. We've done the right thing here. This is going to hurt, but it's not going to be over when they take those trees down. We're told by our elders to stand back and watch. You do what you feel you have to do," he said. "We'll just pray when they come."

In the east, pink streaks appeared between the clouds. A passenger jet roared overhead, so low the name of the airline was legible. At daybreak, the school bell mounted on the camp kitchen began to clang. Drums and whistles sounded under the trees. A line of State Patrol officers in full Darth Vader riot gear marched single file down the hill toward camp with Captain Kevin Kittridge in the lead. The land was about to change hands.

Ceremonies and speeches, then the first wave, all those who didn't want to be arrested, washed up the VA Medical Center hill and clustered a couple of hundred yards away, behind the yellow police line. Below, the figures gathered around the trees looked tiny. People knelt in the dry grass; police began loading them into vans. On the hill, the women ululated: the high familiar wail.

A fire truck hosed down the bonfire. Next came the Hydro-Ax. "Here they come," someone muttered. "Look at the big ugly tree machine."

"I see the courts of the land are totally worthless," said Linda Brown. She wore a stocking cap and a blanket draped around her shoulders, giving her a scraggly, protester look. "This is compliance with the Civil Rights Act? I don't think so. It means nothing, it all means nothing. You call the governor. You call Senator Wellstone. You protest, and you're called a bunch of radical nonworking slobs. But this is the land of the free, got to keep that in mind," she added sarcastically.

Far below her, bulldozers began clearing the camp down to bare earth. Under a gray sky, a cold wind swept smoke from the doused fire across the ground. A Caterpillar with a clawlike hand scooped the kitchen into an overflowing Dumpster. Bales of straw, Styrofoam coolers, splintered boards—it all looked like garbage now. Another bulldozer pushed the Starlodge over.

As had been agreed upon beforehand, Bob Brown supervised it all to make sure ceremonial objects were treated with respect. Seen from the hill, he was a tiny figure in a bright red warm-up jacket. Walking through the piles of debris, escorted by brown-clad state troopers, he looked like the captured officer of a defeated army.

By afternoon, the tipis and sweat lodge were gone, along with the bright wrappings that had swathed the four oaks' trunks. The Starlodge was a pile of trash. Two giant Dumpsters were overflowing. Beneath the oaks, Jim, Bear, Bob Brown, Clyde Bellecourt, and Darlene Jackson, the Ojibwe elder, wandered back and forth, forlorn figures with blankets around their shoulders. They burned sage and sang, the sound floating faintly on the wind. Around them, the big machines seemed to be moving slowly, gracefully in time to the music.

A claw on the end of a mechanical arm lifted a fallen tree and dropped it in a pile of others. A long gray dump truck pulled away, full. Workers were sweeping the field bare, taking everything before the last four oaks. Clyde Bellecourt began a shuffling dance, waving a hawk wing fan above his head. The wind rustled the leaves on the oak trees up on the hill. The yellow police tape fluttered. People stood stiffly, waiting for the end as the wind cut through their clothing, cold as ice water. "Looks like they're saving the four trees for dessert here," someone said.

As dark fell, a worker in a cherry picker hovered in the air, removing four eagle feathers tied in the branches of the four oaks. Then a man with a chain saw stepped up to the first trunk. The crowd on the hill yelled "Murderers!" but their voices sounded shrill. Even they knew it was over. Tarzana and a friend embraced dramatically for a television camera, their eyes cast tragically upward, but the tableau rang false. It seemed, suddenly, that the struggle had been lost a long time ago, although it would have been hard to say just when.

The oaks went down one by one, toppling in slow motion. Their branches hit the ground first, quivered, and crumpled. Their gray wrinkled trunks crashed to earth and lay still.

Watching the oaks fall, Captain Kevin Kittridge thought about the Mendota Dakota. "I felt bad for them," he said later. "I don't know that I believe the four trees were planted there for the reasons they had been told. At the same time, I've been an outdoorsman my whole life, and cutting down a big oak tree isn't good. They're things of beauty. Whether they have another connection or not I don't know, but those were great trees, just gnarly and beautiful oak trees. I was sad to see them dropped."

Natalia and Dr. Toxic saw it all from the top branches of the northernmost oak on the Noble Oaks Trail. At daybreak, they and seven others had climbed into the treetops at the edges of the field, a few hundred yards from the four oaks. "It was just a matter of waiting, being really cold, and watching, tree by tree," Dr. Toxic said. He watched as squad cars surrounded the camp and workers fanned out into the woods. "First they cut down the

Solstice holds forth from one of the few remaining branches of a soon-to-be-felled oak. Photograph copyright 2001 *Star Tribune*/Minneapolis–St. Paul.

trees with nobody in them, then they had enough room to drive vehicles through the woods, 'cause they'd cleared all the trees. They'd go to a tree that a person was in, they'd it cut limb by limb, till it was a *stick*, then they'd remove the person and cut down the tree."

Solstice made quite a scene when they came for him. "The whole time they were cutting parts of his tree down, he was in the guy's face," Dr. Toxic recalled. But it was no use. They got Solstice, and Rory, too.

"We watched them bulldoze the Starlodge, we watched them take down all the brush, we watched them cut down all the other trees. We saw it all," Natalia said.

Dr. Toxic was in a hammock strung from his tree's top branches when they finally came for him. "I didn't even fight. I just let them cut a hole in my net and grab me," he said. "I didn't have any energy to play cat and mouse. I just said, whatever. It's over."

CHAPTER 10

Coldwater Nation

In January a light snowfall covered the field where the camp had been. The squad cars and heavy equipment were gone. A school bus bench sat alone in a flat expanse littered with shreds of plastic and a few broken branches. A straight path through the trees had been shaved clean, as a surgical patient is shaved for the incision. To the south, the dark specks of the cars on Highway 62 whizzed by, now visible from the field. On the snow-covered lookout hill, Marshall Law stood in his old spot, hands clasped behind his back, feet planted wide, gazing to the south.

But when a reporter walked closer and began climbing the lookout hill, the ground slipped underfoot. It wasn't the lookout hill; it was a small mountain of snow-covered wood chips. And up close, the lone sentry proved to be not Marshall Law, but someone named Dan.

Dan was "of the dog-walker crowd, not the protester crowd," he explained. "Funny how every trace of them has been obliterated," he said of the protesters. "I never see them around anymore." From atop the wood chips, Dan surveyed the now protester-less view.

Bundled up to his eyeballs in the bitter cold, Dan wondered

why the protesters were staying away. Maybe the reason was a court order, he said. Still, he was surprised. "Why don't they come back and pitch a couple of tents?" he asked. Dan wondered where they all lived now, with the camp gone. "It's like those cults that say the world is going to end and the spaceships are coming. Then the world doesn't end and what do they do?"

Today Dan was waiting for a fellow dog walker. Soon a woman in a chic red coat and matching hat strode toward him across the field, her German shepherd bounding around her. The three of them set off for the river. "I hope you find some evidence of the protesters," Dan said over his shoulder. "I don't think their efforts were completely wasted." He paused. "Just mostly wasted."

That night, a crowd hugged and laughed and crowded into the old-fashioned wooden pews at Walker Community Church in south Minneapolis. Jim Anderson, Carol Kratz, Bear, Marshall Law, Wes, Solstice—everyone had come for a fund-raiser and send-off. All the camp's occupants seemed to have found new places to live, or at least a friend's floor to sleep on. They were moving on. Marshall and Soil and Wes and Solstice and Emily and around ten others were leaving soon to help the Navajo elders at Big Mountain. Tonight's program would raise money for gas and food for the journey.

It had been a month since the four oaks went down, and the highway fight seemed to have claimed its place, in the minds of many there in the church, as one more chapter in a long history made up of similar struggles against the odds. American Indian Movement cofounder Dennis Banks urged the crowd to help free Leonard Peltier, imprisoned now for almost twenty-four years for his role in a 1975 shootout at the Pine Ridge Indian Reservation in South Dakota in which two FBI agents died. "Write to somebody, anybody, [and] demand that Peltier be released," Banks urged.

Their work at Big Mountain would be a long striving, AIM's Clyde Bellecourt told the young activists. Like the fight to win back the Dakota sacred land of the Black Hills, the resistance at Big Mountain had persevered for many years. "So those of you who are going, take a lot of warm clothes and sleeping bags," he said. "It's going to be cold in the mountains, but I applaud you."

Bellecourt was in poor health and short of breath, his once-lean warrior's face fleshy. The American Indian Movement, in its day as famous as the Black Panthers, had melted into obscurity years ago under the pressures of FBI prosecutions and internal strife. Bellecourt had continued his dogged fight against the injustices suffered by American Indians, but he seemed at times to live in the past. His speech slid quickly into AIM's long struggle and its 1972 march on Washington. "There were over thirty thousand of us," he said. "We have had tremendous impact, but it's because of people like you."

Marshall Law stood onstage and called for a round of applause for Bear, "an inspiration to all of us." He and Solstice dedicated a song to the Mendota Dakota and to Natalia, who was doing twenty days in jail for her part in the lockdown in the State Capitol rotunda.

"Arise, ye children of the Free State," Solstice and Marshall sang. "Listen to the story of the uprising of 1862.... Listen to the story of the uprising of 1998." The words rose, quavering. "Arise you citizens of America, your time has come. It's a war for the water, it's a war for the trees, it's a war against corporate dominion." The audience applauded warmly, like families at a school concert.

Next someone introduced "the legendary Smudge Boys," Jim Anderson and his cousin Michael. ("Smudge" was a reference to the practice of "smudging," or burning sage for spiritual purification. At camp, Jim and Michael had often carried burning sage in a coffee can at the end of a forked stick and cheerfully "smudged" whoever was around.) The Smudge Boys dedicated their song to Carol Kratz and her family: "We may be legendary but we don't practice a lot," Jim joked. "This one's going out to you all." They had composed music to the words Carol and her family had written long ago on the walls of the house on Riverview Road: "Goodbye old house.... My heart will always be here with you." Sitting in his father's lap, Jim's little boy Joey sang along.

When summer came that year, the protesters' garden lay under a twenty-foot-tall earthen berm, part of the massive sound wall that would soon front the remaining houses on Riverview Road. At the end of the street, bleeping dump trucks had taken full possession of the field where the four bur oaks once stood. Lines of cement pillars marched across the precisely leveled earth. The woods on either side of the future roadway looked like a green backdrop: scenery on the margins, the blur that flashes by the car window.

Every week, Paul Eaves, the Wiccan, visited the wood chip labyrinth the witches had made. It was still there at the edge of the field, just outside the orange construction fence. He walked the circular paths and left offerings of lavender. It was a service performed for the dying, "hospice care for the land," he explained. From the woods at the edge of the construction site, a cardinal's whistle rose above the din. "That's Buster," he said. "My father's nickname was Buster. I call all cardinals Buster."

Paul had known all along that falling in love with the land was a risk, but he'd done it anyway. He still thought fondly of his camp name, Blue Moon. The encampment had been a romance between people and the land, he thought—a wondrous time in which the magic of the place became apparent to many. Paul realized now that the message of the trees had been about saving the

Coldwater spring; he had called Senator Paul Wellstone's office about it. Wellstone was aware of the issue and didn't want to see the spring destroyed, Eaves said. The struggle wasn't over yet, by any means. "The land will speak to you if you listen to it," he said. "And what it will say, I have no idea."

Wes, when his train-hopping, dissent-fomenting wanderings took him through town, also visited the site of the camp. "It was the place where I first began to see trees as conscious beings and understand their culture and language and society," he said. The camp had seemed to him like another universe, another dimension. He met witches there for the first time, and learned why so many different people had been drawn to the camp: Paul had called the coven together. The witches did a ceremony and sent out an energy beacon, like a dome with a pulse that spread out in all directions. "One year later to the day, the occupation began. I give them credit for that. They were working with the energy of the place," Wes said. He wasn't surprised that, in the end, the highway's opponents had been defeated. "It would have taken a miracle," he said. "It would have taken five miracles."

At the fringes of the construction, the Noble Oaks Trail still led into the tiny oak forest where Squash and Shane had once strolled and joked beneath the branches. The trail ended abruptly now in bright sunlight and bare earth where the woods had been razed.

In the remaining patch of forest, a few anemones bloomed like stars in the tall grass. Back in the thickets of buckthorn and sumac, a patch of flattened grass marked where a deer had lain. The tiny woods were still a place where a child might come to hide, or to build a fort, or stumble on things hidden in the grass. A wooden pallet lay just off the trail, next to three plastic water jugs tied together with twine and marked in black felt pen "Dr. Fucking Toxic, Minnehaha Free State."

These were the artifacts left by a group of people who watched the world from trees, who looked for inspiration to Druids who danced long ago in sacred groves under the full moon, and to Luddite craftsmen who hurled wooden shoes into the cogs of the industrial revolution. Nearby lay artifacts of another kind: a pile of stakes penciled with a surveyor's hieroglyphics.

On a bright June day, the wind stirred the branches of the weeping willow by the Coldwater spring, and high clouds scudded across the sky. An aging van with dreamcatchers dangling from the rearview mirror and bunches of diamond willow branches tied to the roof pulled into the Bureau of Mines parking lot, and Linda and Bob Brown stepped out. They had come for a pipe ceremony the Mendota Dakota were holding that day at the spring. On her cell phone a minute earlier, Linda Brown had finished working out the arrangements with the authorities. "Sounds like they're going to let everyone drive in!" she said happily. "It always works out the way it's supposed to work, you know."

These days things were going well for the Mendota Dakota. They had their own office now. They and the post office shared a picturesque stone building on Mendota's main street, a block from the Henry Sibley House. Although they were still trying to win federal recognition as a tribe, they had also formed a nonprofit organization dedicated to preserving Dakota heritage and culture. They planned to hold a powwow in Mendota later that summer: the first one, as far as they knew, since 1886. Mendota's mayor had been leery of the idea, but Captain Kevin Kittridge of the State Patrol, the officer who had worked with Jim Anderson and Bear before the final raid, had written the Mendota Dakota a letter of reference. ("I told him I'd dealt with these people for a long time, and they were people of their word," Kittridge said.)

It had fallen to Kittridge to pick up the pipe Jim left under the four oaks. Kittridge had been quite sick afterward, and he couldn't help thinking of Harry Charger's prediction of misfortune. "In the back of your head, you go hummmm ..." Kittridge said. After that, Darlene Jackson, the Ojibwe elder he had gotten to know out at the camp when he smoked the peace pipe, called him often to make sure he and his family were all right. Kittridge wasn't worried, but he appreciated her concern. "Darlene and I are buddies," he said.

Now, beside the spring and the weeping willow, around sixty people stood in the circle while Bob Brown, Jim Anderson, Clyde Bellecourt, and Sky passed the pipe. Billy, a deep-voiced six-footer with purple-lacquered fingernails, a bright pink blouse, and a skirt printed with butterflies, stood with the men inside the circle and gave out pinches of ceremonial tobacco.

"In the time that we fought, you might think that we lost, but we haven't lost anything. The water is still flowing," Jim said. Some in the crowd wore buttons marked with a circle in four sections—red, white, black, and yellow—and the words *Coldwater Nation*. "That's what our prophecies say, that the four colors of man will come together to save the world," he said.

The Mendota Dakota hoped that someday the area around the Coldwater spring could be set aside as an interpretive center where people could learn about Dakota history. Anderson and stalwarts like the red-haired activist Susu Jeffrey sat through endless meetings tying to fend off threats to the spring. They worried that a drainage system for the new interchange to the south would affect the water flow. They had also discovered that the Metropolitan Airports Commission planned to buy the Bureau of Mines campus and build a seven-acre parking lot, so they turned out for public hearings. They sat patiently through the detail-laden proceedings of regulatory boards. "It was one of the strangest meetings I've ever been to," a longtime administrator recalled of one such occasion. About fifty protesters had attended, including a cross-dressing Native American, he said. "There were some very eloquent speakers. The Mendota Dakota are very passionate about this."

That year on May Day, the international workers' holiday, more than four hundred activists took to the streets in Minneapolis. It was the biggest May Day protest the city had seen in years. Rory and Freedom, still together after their time in the cottonwood tree, marched side by side.

A few weeks later, members of Grain Rage, a group opposing genetic engineering, chained themselves in front of the headquarters of Cargill, the Minneapolis-based agribusiness giant. They shut down traffic for miles. After Rory and the rest were cut loose and arrested, two protesters wearing biohazard suits remained on the scene to act as spokespeople. One was Freedom, who spoke into the microphones along with a young man from out of town who declined to give his name. He gave out a contact phone number, though, and to any reporter who had spent much time looking at Minnehaha Free State press releases, the number looked familiar—it was Bob Greenberg's.

Later that summer, word of an event called Stop the Mad Scientists began circulating on the Internet. It would be a "counter-conference and direct action convergence" to confront scientists who were coming to Minneapolis from around the world for a conference of the International Society for Animal Genetics, or ISAG. The Stop the Mad Scientists schedule for Monday, July 24, read "SHUT EM DOWN at the ISAG Conference ... March at Noon ... Autonomous actions city wide!"

The Minneapolis Police Department response to the messages was no surprise to veterans of the Highway 55 fight. The department bought a hundred riot gear suits and announced plans to deploy more than seven hundred officers from the Minneapolis Police Department, the State Patrol, and the Hennepin County Sheriff's Department.

On Friday, July 21, the doors of Walker Community Church in south Minneapolis stood wide open to the warm summer evening. The entranceway was crammed with card tables full of literature and posters ("Biotechnology—The New Answer to Everything from the People Who Brought You Nuclear War"). In the church basement, the counter-conference participants were deep into the scheduled agenda item, "Discussion on SECURITY CULTURE and Update on police mobilization."

Outside, a brightly painted blue school bus loaded with scavenged organic vegetables was parked at the base of the church steps. Tumbleweed lounged in the bus's open door, chatting with friends. His fuzzy beard was gone, replaced with several days' worth of stubble. He wore a red T-shirt with "Stop the Mad Scientists" across the front, and "Resistance Is Fertile" on the back. He stepped out of the bus to talk to the reporters trolling for news on the church steps.

"The T-shirt you're wearing, that's not the name of the group, right?" a newspaper reporter asked. Tumbleweed said it wasn't.

"Do you know if there are any protests around town? It seems pretty mellow," the reporter said.

"Who's to say?" Tumbleweed replied amiably. "There's no leadership, no hierarchy. We've just basically invited people to come and speak their mind."

The reporter, filling in that night from the newspaper's business desk, wanted to know about the bus. She asked another person wearing a "Stop the Mad Scientists" T-shirt.

The bus was called the Coldwater Café, run by a group of folks who started a collective, replied Meaghan, the Sierra Club canvasser of so long ago. The idea had sprung from the Minnehaha Free State, she said.

"What's the Free State?" the reporter wanted to know.

"It was an almost two-year road protest.... The Minnehaha Free State is what *we* like to call the encampment," Meaghan explained patiently. "After the Free State was raided on December 11, 1999, this same group of people has continued to serve breakfast every Sunday morning."

"Are they all ages?" the reporter asked.

"People from all walks of life," Meaghan replied solemnly.

In the demonstrations and police crackdown over the next few days, many of the players were familiar to anyone who had spent time at the Free State. Freedom got her picture in the paper. Emily and Madeline, the seventeen-year-old whose hip was dislocated when she was tackled by the police, were dressed as "radical cheerleaders" in skimpy black skirts and combat boots. They minced and flounced in a dead-on parody of the cheerleaders of their not-so-distant high school days: "Come on girls / let's keep 'em green! / We don't want / designer genes!"

Tumbleweed wore a polo shirt and hid his long hair under a baseball cap. Meaghan lofted a black and green ecoanarchist flag. The genial bearlike figure of veteran Earth First!er and dragon builder Bill Busse strolled through the crowds. It was a nice place to meet people you hadn't seen in a while, he said.

Dr. Toxic stayed away but got picked up anyway, in a vanload of activists two miles distant from the action. After some discussion, the cop let one of the activists go but told Dr. Toxic if he answered *one* more question with a question his ass would be in jail. He did, and it was.

The police response to the ISAG demonstrations cost more than a million dollars, setting a new record. The way Tumbleweed saw it, when you added that to the tab for the Highway 55 raids, a core group of activists had cost the city and state about two million dollars. "Not bad for a bunch of riffraff," he said. Plus, the ISAG demonstrations had been the first in the nation to protest the genetic engineering of animals. It had been a milestone, Tumbleweed thought, just as the Free State had been.

"People ask us, is it over? And yeah, it's over, but we never lost," said Tumbleweed. "We did what we set out to do—we created a community of resistance. There's amazing people that wandered into Minnehaha out of curiosity and are now some of the most kick-ass activists I've met." He didn't like to name names, but he said many people from the Free State had gone to Seattle and to Washington, D.C., for the demonstrations there. He had been in Seattle himself when some fifty thousand people took to the streets to shut down the meeting of the World Trade Organization. "That was the greatest feeling of liberation I've ever felt. A lot of people were born that day," he said. Now in Minneapolis he believed he was seeing a "rebirth of resistance. I like to think a lot of that came out of Minnehaha."

In fall 2000, almost a year after the four oaks were cut, Marshall Law was in Oregon studying herbal medicine. Nettle was in Vermont. There were trees there that needed defending, she figured, since the timber companies had cut their way from the East Coast to the West, and were now casting their eyes back east. Emily was on the road a lot, back in town mostly for court appearances. She walked with a long, loping gait now, in heavy hiking boots and low-slung black Carhartt work pants with a Leatherman, the Earth First!er's all-purpose action tool, at her belt. Solstice worked at the Seward Café ("collectively owned since 1974") in south Minneapolis. These days he thought a lot about how activists might better connect with the communities they were trying to serve. Tarzana had changed her name back to Anne-Marie and had left town. Dr. Toxic and Natalia were expecting a baby—a future activist, they hoped—and working to save money, he as a dishwasher, she as a receptionist.

Captain Kevin Kittridge and Lieutenant Bud Emerson had both been promoted. Emerson was still trying to make a difference in the world. Now he worked in the sex crimes unit of the Minneapolis Police Department. He respected the activists for their efforts to make a difference. "Unless we have activists we won't have any change," he said. "It's like Machiavelli says, anyone who intends to change society will have—now to paraphrase—only the lukewarm support of those who might benefit from the change, and the undying enmity of those who benefit from the status quo."

Tumbleweed was still as fervent a resister as ever, but that fall the strain was starting to show. The pages of his anarchist Slingshot Collective Organizer daybook were marked with court hearings. Unlike the vast majority of protesters who had been arrested, whose charges amounted mostly to misdemeanors, Tumbleweed faced serious jail time. He had been arrested in July 2000, just after the ISAG counter-conference, during a police raid on a house filled with anti-ISAG activists. His billfold lay on the mantel of a fireplace in the same room as a desk in which police found a small plastic bag of hallucinogenic mushrooms. He was charged with possession of the mushrooms, a felony. Besides the criminal charges, Tumbleweed's other souvenir of the raid had been an oozing red and purple shiner that swelled his right eye completely shut.

The day of his pretrial hearing that fall, Tumbleweed waited outside the courtroom, in a hallway high in Hennepin County's towering steel and glass government building. He had taken a shower and gotten dressed up for the occasion, but the result was not impressive. He wore a ratty tie, a pair of worn hiking boots, and a thrift-store suit—approximately his size—that somebody had left in the house where he was staying. "Oh my God, Tumble in a suit," someone remarked.

Tumbleweed had dark circles under his eyes but maintained his usual bouncy energy. The authorities were suffering from "post-Seattle stress syndrome," he said. "The system is on the verge of collapse. Those in power are worried or they wouldn't be doing what they're doing."

A crowd of Tumbleweed's friends and supporters, some forty-five people, had come for the hearing. Dr. Toxic leaned against the glass wall of the government center's atrium. He regarded floor after floor of office cubicles, each occupied by busy government workers, like bees in a honeycomb. "I've always wanted to do a banner hanging here," he said cheerfully. "Not that I ever would."

The courtroom doors opened and everyone filed in for the hearing. They looked different, as they always did, from "normal people," as they sometimes called the mainstream world. "Normal people" lived self-centered lives, spent their free time going to shopping malls and watching television, and wondered why their families and communities were falling apart. The Free State people lived in a different world—a world, whatever its short-comings, they had built for themselves. Now they had come to support one of their own. Before the charges against Tumbleweed were finally dropped, the hearing would be the first of many as his case dragged on. His friends would be back for each one.

"We have an amazing community that came out of the Free State still. I think about people who don't have communities like ours and I wonder how they survive," Natalia said. The fall before, when the camp still existed, she'd had a miscarriage and been hospitalized for five days. "I walked around every once in a while and looked at all the people in their hospital beds alone, with nobody there. And then I came back to my room, and there were like ten people hanging out. When I went to jail, in 2000, I got support from all kinds of people. I got a pile of mail this big. That doesn't happen to a lot of people, and it happened to me because I had this amazing community that I'm surrounded with. I don't think anyone who was even briefly exposed to the Free State will ever be the same again. I don't think people can just go on with their lives as they were."

"It was awesome. It was beautiful," Dr. Toxic said. "I'd never ever experienced that. And I want to experience that more and more."

Natalia and Dr. Toxic had a baby girl on December 30, 2000. They named her Nara Joy Coldwater.

On a gray November day that same winter of 2000, Bob Greenberg strode through the Minneapolis–St. Paul airport, snared an unattended cart, and loaded it with his purple high-tech backpack, his guitar, and a cardboard carton stuffed with newsletters and copies of *Green Anarchy* and the *Earth First! Journal*. He had a job as an organizer for an international network of forest activists. He was going to Fiji by way of New Zealand.

Bob's first stop would be a conference of "international trade activists, forest defenders, and indigenous people from every continent" in New Zealand. His experience at Minnehaha had helped him learn to work with indigenous people, he felt. "For me personally, one of the most important things I've ever done in my life is going to the Mendota people and getting them involved," he said. "It's one of the most meaningful things I've ever done."

One of his goals these days was to develop support campaigns for indigenous people who were taking up arms to defend their land. "Well, you know the Zapatistas took up arms, there's some support campaigns for that," he said. "The Miskitos on the eastern coast of Nicaragua are taking up arms, just last week you know, to protect the area down there. The Ucizoni, which is an alliance of a hundred different tribes from fifteen different nations in southeastern Mexico adjacent to where the Zapatistas are, they're considering taking up arms to stop the dry canal megaproject."

On the plane, Bob planned to use his laptop computer to outline a strategy. It involved using a sophisticated computerized mapping and analysis system known as GIS—Geographic Information Systems—to identify areas of native forest where indigenous people still lived in balance with the earth. "Basically, the proposal I'm making is to do a GIS map, draw lines around this area, and then come up jointly with a strategy, with each group playing the role that they're most comfortable with," he explained. "Whether it's the indigenous people saying, 'Hey, we're going to defend our land and our lives, and we're going to take up arms.' And then other groups say, 'OK, we'll do media support for you,' and other groups say, 'We'll send down human rights observers,' and others saying, 'We're going to challenge this in international court, or U.S. court.' So whatever each group feels comfortable with, finding a way that we can plug jointly into strategies."

Bob strode by an eight-foot-tall inflated statute of Ronald McDonald and gave a surreptitious tug on its plug, which held fast. He sat down in the boarding area with a reporter and answered a few questions about the Minnehaha campaign. Was there anything he would do differently if he had it to do over?

He would have increased the level of direct action, he replied. "I also would have increased the level of covert actions that happened, and encouraged a much higher level of property destruction. More bulldozers being decommissioned, more damage being done to the plants at McCrossan and to the equipment at McCrossan. Really, just more destruction while the road was being built. Damage to the road itself, so that they would have to do road repair and continue to spend money. But really driving up the cost, you know, and the stakes in that sense.

"Personally, in the position I'm in now, I would advocate armed struggle under the situation that we were in. You know, when it came down to the last point, the last December 11 in the trees, people shouldn't have just done a symbolic struggle and gone away. We should have stood our ground. We should have barricaded ourselves in the Bureau of Mines. Armed. And defended the area. We can't allow this to keep continuing. We have to draw a solid line. And there's a difference between being violent and engaging in self-defense."

Bob's flight for New Zealand was boarding, but he kept talking, louder and more passionately. "We are killing the planet, and at some point she's not going to die, she's going to shake our species and a lot of other species off like a bad set of fleas and heal herself. The survival of our species depends on us defending what it takes for us to survive. If it comes down to armed struggle, then that's what it takes."

Loudspeakers blared a last call, and Bob finally stood up. His face lightened. "I'm excited about going to Fiji," he said. "There's a revolution going on there. Who knows? Maybe I'll be eaten." He smiled and bounded toward the plane.

Wes continued to foment dissent in various places, always traveling light. He came and went on his mysterious wanderings. He didn't have an e-mail address ("Certainly not," he replied good-naturedly when he was asked), but with luck a person might run across him at the annual Big Woods Earth First! Rendezvous at a park in southern Minnesota, say, or, getting a free meal after hours at the Seward Café before hopping the next freight out. It was just a matter of knowing where to look. "I'm not really that hard to find if you come where I am," he said.

Finding Jim Anderson was easy. Every Monday afternoon at around 2:15, a van with a "Born Again Savage" sticker would cruise past the steel gates of the Bureau of Mines campus and park by the Coldwater spring.

Jim would kneel by the weeping willow, by the side of the pool fed by the spring, and open the wooden box that held his pipe, his sage, and his tobacco. He would burn some sage to purify the air, hold up a pinch of tobacco and pray to the four directions, then pass the pipe to whoever had come that day for the ceremony. Afterward, they would discuss the quest to preserve the spring.

On a drizzling November day, Jim passed the pipe to some people called John and Marigold and Marigold's dreadlocked friend from out of town. Jim said he had seen a bald eagle recently, at a ceremony. It had alighted on a tree nearby. "It was Grand-father come to show us that we're doing the right thing," he said, his voice thick with emotion. "Those kinds of things keep you humping and pumping on this. It's proof." Jim talked about calling Senator Wellstone's office for a letter of support. Things seemed to be going well.

"We've got the parking lot whupped," said John.

Jim hoped to get Minnesota's federally recognized tribes to buy the Bureau of Mines land for an interpretive center. "They could buy it in a heartbeat," he said. "It's a perfect opportunity. I haven't got this all on paper like I should have, but that's my thought on it."

Rain mixed with snow made circles on the pool's surface. Drops fell with a faint tapping on the still-green grass, the fallen

yellow leaves of the weeping willow, and the sad, decaying buildings of the Bureau of Mines. "It's working. We just gotta keep
going," Jim said. "We prayed a lot at the camp. It's going to work
out. We're going to win."

Jim himself didn't need any convincing. Harry Charger had
told them all along that they would lose the trees but save the
spring, and it had all worked out, just as he said.

Afterword

On May 15, 2001, Governor Jesse Ventura signed into law a bill protecting the Camp Coldwater Spring from any state action "that may diminish the flow of water to or from" the spring. At the time, few lawmakers expected it to have any effect on the Highway 55 project.

But a month later, dye testing commissioned by the local watershed board showed that construction of the project's final phase, the intersection of Highways 55 and 62, was cutting the spring's flow. The Minnesota Department of Transportation tried and failed to get the interchange exempted from the law. That September, the agency announced it was halting work on the interchange and could not finish it unless the law was repealed.

Repeal efforts in the 2002 legislative session failed. At this writing MnDOT and the Minnehaha Creek Watershed District have agreed on a new plan for the intersection, designed to keep the spring's flow intact. The Mendota Mdwakanton Dakota still hope to have the land around the Camp Coldwater Spring preserved as a Native American cultural center.

April 2002

Acknowledgments

I'd like to thank everyone who helped me with this book, including Peter Leschak, who convinced me I could write a book in the first place; my editor, Todd Orjala, who believed in the story; my writers' group, who offered advice and encouragement during wobbly first drafts; and the many friends who read the book and offered suggestions. Special thanks to Catherine Winter, Chris Julin, and Liese Greensfelder, whose friendship and steadfast guidance saw me through.

Note on Sources

Sources for the histories of the Dokota conflict and the Mendota Mdewakanton Dakota Community include: *Through Dakota Eyes: Narrative Accounts of the Minnesota Indian War of 1862*, edited by Gary Clayton Anderson and Alan R. Woolworth; *Bury My Heart at Wounded Knee: an Indian History of the American West* by Dee Alexander Brown; *Little Crow, Spokesman for the Sioux*, by Gary Clayton Anderson; *Kinsmen of Another Kind: Dakota-White Relations in the Upper Mississippi Valley, 1650–1862*, by Gary Clayton Anderson; and *History of the Santee Sioux: United States Indian Policy on Trial*, by Roy W. Meyer.

Many of the nineteenth-century newspaper accounts quoted in the book were unearthed by members of the Mendota Dakota Community during their research and provided to me by Lisa Krahn, manager of the Sibley House Historic Site. My thanks also to Alan Woolworth of the Minnesota Historical Society for his advice and guidance, and to Minnesota Public Radio researcher Betsy Cole for her help in retrieving contemporary newspaper accounts of Earth First! The section on Earth First! also includes information from Susan's Zakin's *Coyotes and Town Dogs: Earth First and the Environmental Movement*.

Mary Losure is an award-winning environmental reporter for Minnesota Public Radio. She is a longtime contributor to National Public Radio, where her work has aired on *Morning Edition* and *All Things Considered*. She has written and produced feature stories on topics as diverse as Tibetan immigrants in Minnesota, speaking in tongues, and the worldwide decline of amphibians.